Lecture Notes in Mechanical Engineering

Lecture Notes in Mechanical Engineering (LNME) publishes the latest developments in Mechanical Engineering—quickly, informally and with high quality. Original research reported in proceedings and post-proceedings represents the core of LNME. Volumes published in LNME embrace all aspects, subfields and new challenges of mechanical engineering. Topics in the series include:

- Engineering Design
- Machinery and Machine Elements
- Mechanical Structures and Stress Analysis
- Automotive Engineering
- Engine Technology
- Aerospace Technology and Astronautics
- Nanotechnology and Microengineering
- Control, Robotics, Mechatronics
- MEMS
- Theoretical and Applied Mechanics
- Dynamical Systems, Control
- Fluid Mechanics
- Engineering Thermodynamics, Heat and Mass Transfer
- Manufacturing
- Precision Engineering, Instrumentation, Measurement
- Materials Engineering
- Tribology and Surface Technology

To submit a proposal or request further information, please contact the Springer Editor of your location:

China: Ms. Ella Zhang at ella.zhang@springer.com
India: Priya Vyas at priya.vyas@springer.com
Rest of Asia, Australia, New Zealand: Swati Meherishi at swati.meherishi@springer.com
All other countries: Dr. Leontina Di Cecco at Leontina.dicecco@springer.com

To submit a proposal for a monograph, please check our Springer Tracts in Mechanical Engineering at http://www.springer.com/series/11693 or contact Leontina.dicecco@springer.com

Indexed by SCOPUS. All books published in the series are submitted for consideration in Web of Science.

More information about this series at http://www.springer.com/series/11236

Bangarubabu Popuri · Amit Tyagi · N. R. Chauhan ·
Ashish Gupta
Editors

Recent Trends in Engineering Design

Select Proceedings of ICCEMME 2021

Springer

Editors
Bangarubabu Popuri
Department of Mechanical Engineering
National Institute of Technology, Warangal
Hanamkonda, Telangana, India

Amit Tyagi
Department of Mechanical Engineering
Indian Institute of Technology BHU
Varanasi, Uttar Pradesh, India

N. R. Chauhan
Department of Mechanical
and Automation Engineering
Indira Gandhi Delhi Technical
University for Women
Kashmere Gate, Delhi, India

Ashish Gupta
Department of Mechanical Engineering
GL Bajaj Institute of Technology
and Management (GLBITM)
Greater Noida, Uttar Pradesh, India

ISSN 2195-4356 ISSN 2195-4364 (electronic)
Lecture Notes in Mechanical Engineering
ISBN 978-981-16-2899-3 ISBN 978-981-16-2900-6 (eBook)
https://doi.org/10.1007/978-981-16-2900-6

This Springer imprint is published by the registered company Springer Nature Singapore Pte Ltd.
The registered company address is: 152 Beach Road, #21-01/04 Gateway East, Singapore 189721, Singapore

Preface

International Conference on Computational and Experimental Methods in Mechanical Engineering (**ICCEMME-2021**) has been the third conference of its series organized by the Department of Mechanical Engineering of GL Bajaj Institute of Technology and Management, Greater Noida, Uttar Pradesh, India. The institute is located in the vicinity of the industrial hub. Therefore, it was decided to provide a forum to bring together scientists, speakers from industries, university professors, graduate students, and mechanical engineers, presenting new research in science, technology, and engineering.

The motive of this conference was to provide an opportunity to share their innovative ideas in the form of a paper presentation. Research articles were based on Design Engineering areas such as modelling and simulation, application of modelling to complex real-world systems, application of machine or deep learning in mechanical problems, artificial intelligence, vehicle design, Robotics, vehicle dynamics and control, biomechanics, and vibration-related problems. During the conference, about 10 delegates joined from various countries and delivered a keynote lecture on the theme of the conference. All papers were critically reviewed by two reviewers from national/international authors.

Furthermore, we, ICCEMME, would like to extend our appreciation to all the authors for contributing their valuable research to the conference. The committee is also grateful to all reviewers who spared their time to carefully review all the assigned research articles and to all the committee members for their great effort in making this conference a great success.

We are thankful to all sponsored agencies who gave us their cooperation and funding support.

We are thankful to our management and director of GL Bajaj Institute of Technology and Management, Greater Noida, Uttar Pradesh, India, for their continuous source of inspiration and valuable support. We are thankful to all the members of the organizing committee for their contribution in organizing the conference. Last but not least, we thank Springer for its professional assistance and particularly Ms. Priya Vyas and Ms. Sushmitha Shanmuga Sundaram who supported this publication.

Hanamkonda, India — Bangarubabu Popuri
Varanasi, India — Amit Tyagi
Delhi, India — N. R. Chauhan
Greater Noida, India — Ashish Gupta

Contents

Bond Graph Analysis of Dynamic Interaction Between the Concrete Slab and Subgrade for High-Speed Track 1
Saurabh Bhardwaj, Yamika Patel, and Vikas Rastogi

Identification of Damping in Sandwich Beam 13
Harshit Bahri, Kaushalendra Kumar Singh, and Harvendra Singh

Effect of Groove Dimensions on Pressure Profile of Twin Axial Groove Hybrid Journal Bearing 21
Anamika Yadav and Pooja Pathak

Method of Obtaining Planar State of Stress Using Mohr's Circle—Some Typical Cases 33
Shailendra Singh Chauhan and Avadhesh Kumar Khare

Computational Modeling of Bubble Formation on Submerged Orifice 39
Sachin Kumar and Raj Kumar Singh

Design and Development of a Multivariant Exercise Machine (MEM) for the Patients Suffering from Spine-Related Problems 53
Ruchika Gupta, Sharad Prateek Singh, and Vinod Kumar Yadav

Parametric Study and Optimization in Deep Drawing of Headlight Reflector 67
Jitendra Kumar Singh, Jasleen Kaur, and Rashmi Mishra

Numerical Simulation of a Race Car Wing Operating in the Wake of Leading Vehicle with Varying Diffuser Angles 75
C. Animesh Bharadwaj, V. V. S. Ganesh, and B. T. Kannan

A Review on Design Optimization of Leaf Spring 93
Ranjeet Kumar Singh and Vikas Rastogi

Numerical Study of Twisted Tape with Circular Cutout and Triangular Cutout in a Circular Tube 109
Awasthi Aditya Bachchan, Sourabh Gahlot, Gopal Nandan, Satish Kumar, and Ramakant Shrivastava

Editors and Contributors

About the Editors

Dr. Bangarubabu Popuri is Professor in the Mechanical Engineering Department of the National Institute of Technology, Warangal - 506004, Telangana, INDIA, Dr. Popuri received his Ph.D. degree in Flexible Multi body Dynamics Applied to Robot Manipulators (Mechanical Engineering) in 1992 from IIT Madras; M.Tech. Degree in Mechanical Engineering (Design Elective) from IISc, Bangalore, in 1988; B.Tech. Degree in Mechanical Engineering from Nagarjuna University in 1985. Dr. Popuri has more than 20 years of experience in academics and industry. He has authored more than 10 research papers published in reputed international journals and conference proceedings. He is a member of several national and international advisory committees. He has handled several sponsored research projects with funding support from DRDO, NTPC, etc. He has worked on several industrial consultancy projects. He has supervised a number of M.Tech dissertations and 2 Ph.D. theses. Currently, 8 research scholars are working in his research group. Dr. Popuri is actively engaged in research work on the mechanical vibrations, vibration control, finite element analysis, mechanism science, engineering design, rotor dynamics, vehicle dynamics.

Dr. Amit Tyagi is currently working as an Associate Professor at the Department of Mechanical Engineering, Indian Institute of Technology (BHU) Varanasi. He has 18 years of academic experience. He obtained his B. Tech. in Mechanical Engineering from G. B. Pant University of Agri and Technology, Pantnagar, Uttarakhand. He obtained M. Tech. and Ph.D. from Indian Institute of Technology Roorkee and Indian Institute of Technology (BHU) Varanasi, respectively. His major areas of research interests include vibration analysis, structural health monitoring, FEM and wavelets. He has published 19 research papers in various international journals of repute.

Dr. N. R. Chauhan is working as an Associate Professor & Training and Placement Officer (TPO) in the Department of MAE at Indira Gandhi Delhi Technical University for Women (IGDTUW) Delhi (India). He received his Ph.D. degree in Mechanical Engineering from IIT Roorkee and M.Tech with specialization in Machine Design

Engineering from IIT Roorkee. He did B.Tech in Mechanical Engineering from BIET Jhansi (An Autonomous Institute Funded by the U.P. Government). He headed the Department of Mechanical & Automation Engineering from December 2014 to December 2017. He has also worked as Course Coordinator for M.Tech in Robotics and Automation from October 2014 to July 2019. He has delivered several expert lectures in various faculty development programmes. His students have participated in various car design and fabrication events within and outside of India and won many awards. He was the member of various committees such as board of studies, academic council, financial committee, departmental research committee and others. He is a Life Member of Tribology Society of India and was member of Society of Automotive Engineers, India. He is currently supervising Ph.D students in the field of machine design engineering, tribology, fluid film bearings, alternate fuels and composite materials. He has a teaching experience of more than 18 years. He has published more than 100 papers in international/national journals and conferences of repute.

Dr. Ashish Gupta is currently working as an Associate Professor at the Department of Mechanical Engineering, G. L Bajaj Institute of Technology and Management, Greater Noida, Uttar Pradesh, India. He obtained his B.Tech (Mechanical) from Uttar Pradesh Technical University; M.Tech (Mechanical Engineering) from Sant Longowal Institute of Technology (SLIET Longowal) and PhD from Delhi Technological University (formerly called Delhi College of Engineering), Delhi. His major areas of research interests include vehicle dynamics, bondgraph modeling, system dynamics and control, rotor dynamics, and bio-mechanics. He has published more than 15 papers in reputed international journals and conferences. He is actively involved in industrial consultancy and projects. He has reviewed many papers for different reputed international journals.

Contributors

C. Animesh Bharadwaj Department of Aerospace Engineering, SRM Institute of Science and Technology, Chennai, Tamil Nadu, India

Awasthi Aditya Bachchan Amity University Uttar Pradesh, Noida, India

Harshit Bahri Department of Mechanical Engineering, GL Bajaj Institute of Technology and Management, Greater Noida, UP, India

Saurabh Bhardwaj Delhi Technological University, Delhi, India

Shailendra Singh Chauhan GL Bajaj Institute of Technology and Management, Greater Noida, Gautam Buddha Nagar, UP, India

Sourabh Gahlot National Institute of Technology, Jamshedpur, India

V. V. S. Ganesh Department of Aerospace Engineering, SRM Institute of Science and Technology, Chennai, Tamil Nadu, India

Ruchika Gupta GL Bajaj Institute of Technology and Management, Greater Noida, Uttar Pradesh, India

B. T. Kannan Department of Aerospace Engineering, SRM Institute of Science and Technology, Chennai, Tamil Nadu, India

Jasleen Kaur National Institute of Technical Teachers' Training and Research, Chandigarh, Chandigarh, India

Avadhesh Kumar Khare Quantum University, Roorkee, India

Sachin Kumar Department of Mechanical Engineering, Delhi Technological University, Delhi, India

Satish Kumar National Institute of Technology, Jamshedpur, India

Rashmi Mishra G.L. Bajaj Institute of Technology & Management, Greater Noida, India

Gopal Nandan Government College of Engineering, Karad, India

Yamika Patel Delhi Technological University, Delhi, India

Pooja Pathak Department of Mathematics, GLA University, Mathura, UP, India

Vikas Rastogi Department of Mechanical Engineering, Delhi Technological University, Delhi, India

Ramakant Shrivastava Government College of Engineering, Karad, India

Harvendra Singh Department of Mechanical Engineering, GL Bajaj Institute of Technology and Management, Greater Noida, UP, India

Jitendra Kumar Singh G.L. Bajaj Institute of Technology & Management, Greater Noida, India

Kaushalendra Kumar Singh Department of Mechanical Engineering, GL Bajaj Institute of Technology and Management, Greater Noida, UP, India

Raj Kumar Singh Department of Mechanical Engineering, Delhi Technological University, Delhi, India

Ranjeet Kumar Singh Department of Mechanical Engineering, Delhi Technological University, Delhi, India;
Department of Mechanical Engineering, G.L. Bajaj Institute of Technology and Management, Greater Noida, India

Sharad Prateek Singh GL Bajaj Institute of Technology and Management, Greater Noida, Uttar Pradesh, India

Anamika Yadav Department of Mathematics, GLA University, Mathura, UP, India

Vinod Kumar Yadav GL Bajaj Institute of Technology and Management, Greater Noida, Uttar Pradesh, India

Bond Graph Analysis of Dynamic Interaction Between the Concrete Slab and Subgrade for High-Speed Track

Saurabh Bhardwaj, Yamika Patel, and Vikas Rastogi

1 Introduction

The railway train running along a track comprises of various degrees of freedom as it comprises of many parts. It is one of the most complicated dynamic systems in engineering and the most important mode of public transport in a country like India where it runs more than 20,000 passenger trains daily. All over the world, the railway industries are making considerable efforts to increase the speeds and load carrying capacity of trains. Karmakar and Mukherjee [1] modelled Electric Overhead Travelling (EOT) cranes subjected to severe dynamic loading, using bond graph technique. Zeng and Dai [2] described the bond graph modelling technique for vehicle–bridge coupling system. The track was simplified as a lumped parameter system and the effective rail and sleeper masses were included in the vehicle-bridge coupling model [3].

Nielsen et al. [3] investigated the vertical dynamic behaviour for a railway bogie moving on a rail which was discretely supported\via rail pads\ by sleepers resting on an elastic foundation. Application examples were given in which the influences of three types of practically important imperfections in the compound vehicle/track system were investigated. The third imperfection was a case where a single sleeper had lost its support due to erosion of the ballast. The dynamic of vertical interaction between a moving rigid wheel and a flexible railway track has been investigated by Bureika and Subačius [4]. Transient bending tensions in sleepers and rail were calculated. Kumar and Sujath [5] presented a numerical simulation of the lateral dynamic behaviour of a computed using 17 dof model. Complete model of a railway vehicle,

S. Bhardwaj (✉) · Y. Patel · V. Rastogi
Delhi Technological University, Delhi 110042, India

Y. Patel
e-mail: yamika_phd2k17@dtu.ac.in

V. Rastogi
e-mail: rastogivikas@gmail.com

B. Popuri et al. (eds.), *Recent Trends in Engineering Design*, Lecture Notes in Mechanical Engineering, https://doi.org/10.1007/978-981-16-2900-6_1

which has been assembled from wheel sets, car body and intermediate structure which are flexible, and which have been connected by components such as springs and dampers as considered by Kumar and Rastogi [6]. A combined vehicle-track bond graph model was developed to study the wheel/track impact by Kumar and Rastogi [7]. It was noticed from simulation results that a linear track model was found to be inappropriate for wheel/track interaction problems due to the extremely large value of impact load.

The objective of the present study is to create a three-dimensional bond graph model of the concrete slab and subgrade system, simulation of vertical dynamics of the high-speed railway track system, and to study the effect of vertical dynamics by varying the operational speed of the high-speed train.

2 The Proposed System

To construct a modal model for the concrete slab–subgrade system, for high-speed tracks, the beam (concrete slab) was divided into five modes. The contact between concrete slab and subgrade was considered as a spring–damper system, where the spring and damper represented stiffness and damping properties, respectively. The contact between the moving load and the concrete slab is modelled in a similar fashion. The two concrete slabs under the two parallel rails, connected with cement–asphalt mortar, are modelled in a way that the same modes of both slabs are connected with the help of a spring–damper system, representing the properties of cement–asphalt mortar (Fig. 1).

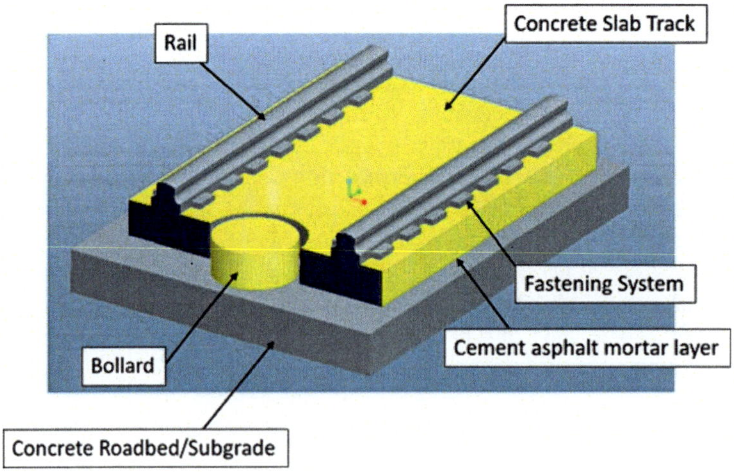

Fig. 1 Physical model of the concrete slab and subgrade system

Equations (3), (4), (6), and (7) are used to write expressions for the bond graph elements. After that, the bond graph was checked for errors and warning, and then the equations were generated successfully. The bond graph system was then compiled and simulated.

3 Numerical Modelling

In this paper, the track, i.e., the concrete slab, is modelled as Euler–Bernoulli beam (Fig. 2).

If the transverse deflection of the beam is taken as z_s (x, t), where x denotes the longitudinal position of the beam, then the partial differential equation of the beam can be derived and given as shown in Eq. (1):

$$EI_s \frac{\partial^4 z_{s(x,t)}}{\partial x^4} + \rho_r A_r \frac{\partial^2 z_{s(x,t)}}{\partial t^2} = P_c(t)\delta(x - x_w) \tag{1}$$

Here,

EI_s—Vertical bending stiffness of concrete slab.

ρ_s—Density of concrete slab.

A_s—Area of cross section of the concrete slab.

$P_c(t)$—Contact force at the wheel/rail/slab interface.

x—The longitudinal position of the concrete slab with respect to the left end support of the beam.

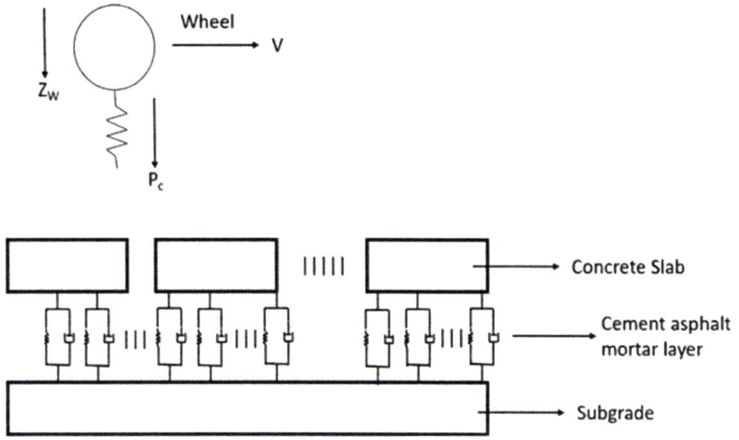

Fig. 2 Computational model of concrete slab and subgrade system

x_w—The position of the wheel.

δ(.)—Dirac's delta function.

The partial differential equation is converted into normal derivatives by the variable-separation method. The solution by this method comes out to be an infinite sum of the product of shape function $Y_i(x)$ and modal amplitude $q_i(t)$, which represents the vertical motion of the slab. In the present analysis, in total 5 modes are considered. The displacement of the rail beam at x from the beam end at time t is given as shown in Eq. (2):

$$z_r(x,t) = \sum_{i=1}^{NM} Y_i(x) q_i(t) \tag{2}$$

where *NM* is the number of modes.

Both the ends of the concrete slab beam are pivoted; therefore, at both the ends, the deflections and the bending moments are equal to zero. Applying these boundary conditions, shape function and natural frequencies for ith mode are obtained, as given by Eqs. (3) and (4), respectively [8].

$$Y_i(x) = \sin\left(\frac{i\pi x}{L_s}\right) \tag{3}$$

$$\omega_i = \left(\frac{i\pi x}{L_s}\right)^2 \sqrt{\frac{EI_s}{\rho_s A_s}} \tag{4}$$

where

Ls is the length of the concrete slab.

Substituting Eqs. (3)–(4) and integrating with respect to x along the length of beam, considering the orthogonality property of shape functions, one obtains Eq. (5).

$$m_i \ddot{q}_i + k_i q_i = P_c(t) Y_i(x_w) \tag{5}$$

where m_i is the modal mass and k_i is the *i*th modal stiffness.

$$m_i = \frac{\rho_s A_s L_s}{2} \tag{6}$$

$$i = 1, 2 \ldots .NM$$

$$k_i = m_i \omega_i^2 \tag{7}$$

$$i = 1, 2 \ldots .NM$$

The modal momentum is given by the expression:

$$p_i = m_i \dot{q}_i \tag{8}$$

Therefore, the Eq. (5) can be written as:

$$\frac{dp_i}{dt} = -k_i q_i + P_c(t) Y_i(x_w) \tag{9}$$

and

$$\frac{dq_i}{dt} = \frac{p_i}{q_i} \tag{10}$$

This reduces the equation, which is an ith modal equation, in the form of q_i and p_i, into two first-order state equations.

Equation (3) is used in the expression of transformer element in the bond graph. Similarly, Eqs. (6) and (7) are used in the expression of stiffness properties of the concrete slab.

4 Bond Graph Modelling

The resulting modal bond graph model for the concrete slab/subgrade system is as shown in Fig. 3.

Here, C172, C164, C28, C35, and C31 represent the stiffness properties of the subgrade. R173, R1655, R29, R36, and R32 represent the damping properties of the subgrade. The 1-elements, joined to the 0-elements by transfer functions, represent the nodes of the beam (concrete slab).

When the bond graph shown above, for a single rail, is to be converted in 3-D, a system of springs and dampers denoting the cement–asphalt mortar is used to join the two rails to model the whole track. The bond graph model for the same is shown below.

It is a three-dimensional modal model of the concrete slab–sub grade system, subjected to the moving load of the train. The concrete slab is divided into five modes.

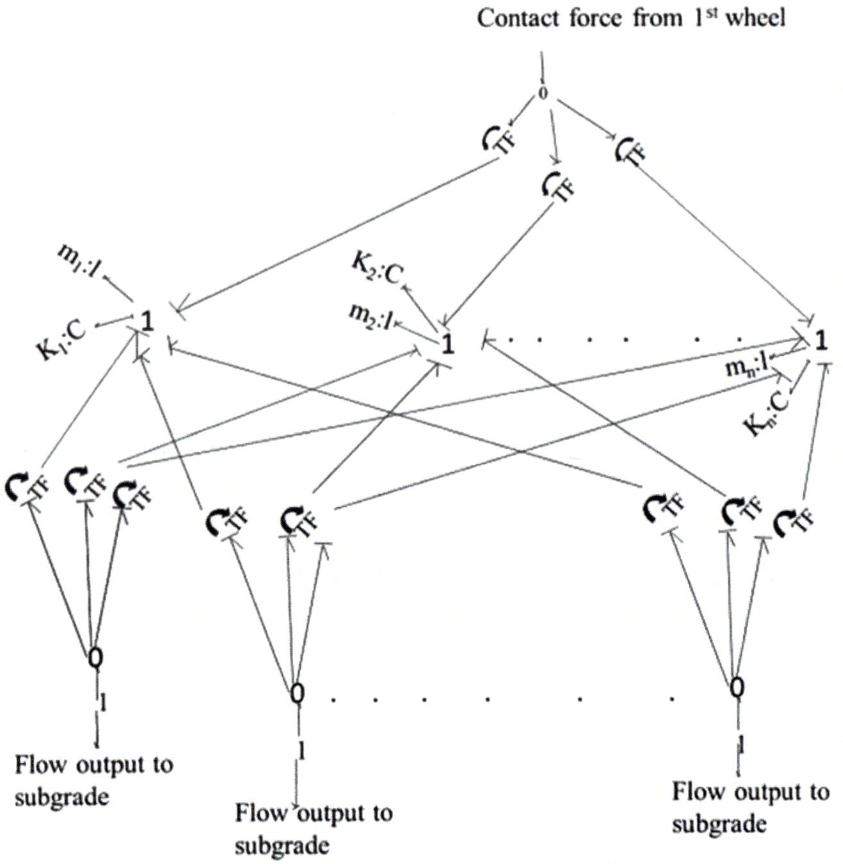

Fig. 3 A representative Bond graph model of the concrete slab and subgrade system

5 Simulation Study

The results obtained, in the form of graphs, are shown in this section, for the simulation of the bond graph shown in Fig. 4. The interpretation of the results is discussed in the next chapter.

After defining all the expressions in the bond graph model and after the generation of all the equations successfully, the bond graph is then simulated. All the values of parameters are to be entered for the generation of graphs in the simulation window, and an appropriate time duration is chosen for the same.

The values of all the properties of the elements, used in the expressions and in the final calculations of the model, during the simulation are shown in Table 1.

The parameters mentioned above are used during the simulation of the final bond graph. The graphs generated henceforth can be altered by varying the values of

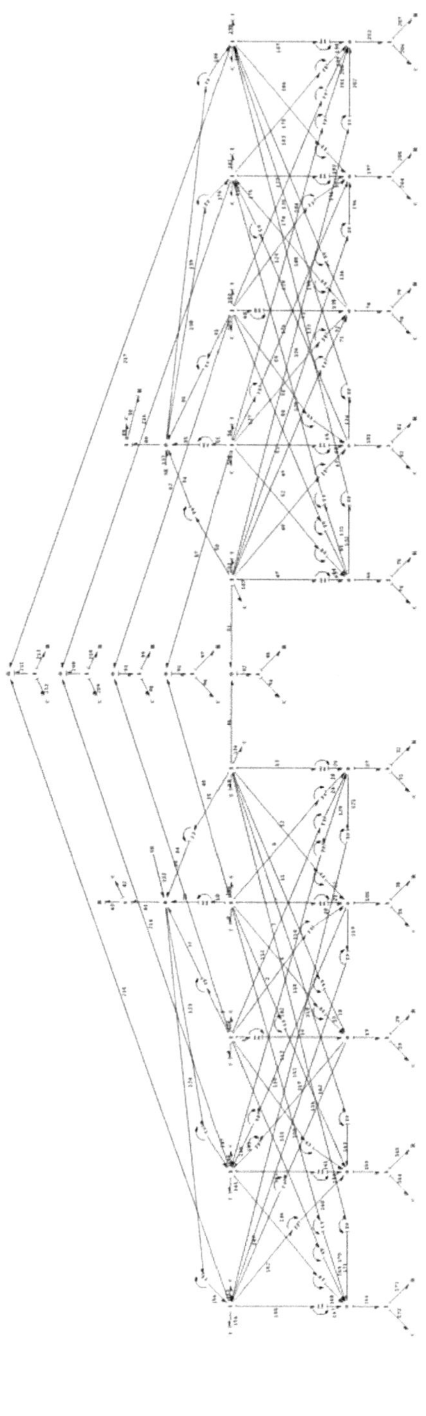

Fig. 4 Three-dimensional Bond graph modal model of the concrete slab and subgrade system, subjected to a constant moving load

Table 1 Material properties of the components of high-speed track

Properties	Values
Stiffness properties of subgrade (C_sub)	60×10^6 N/m
Damping properties of subgrade (R_sub)	90×10^3 Ns/m
Density of concrete slab (ρ)	2500 kg/m^3
Length of concrete slab (L)	15 m
Area of cross section of a single concrete slab (A)	0.435 m^2
Stiffness properties of concrete slab (C_contact)	87×10^9 N/m
Stiffness properties of cement–asphalt mortar (C_lat)	90×10^7 N/m
Damping properties of cement–asphalt mortar (R_lat)	83×10^3 Ns/m
Flexural rigidity of concrete slab (EI)	13.25×10^6 Nm^2
Static load acting on the beam (concrete slab)	80×10^3 N

parameters in the "parameters" section in the simulation window. The trend and the time of operation for the system to become stable, or tend towards becoming stable, is the primary concern.

At various operational speeds, ranging from 200 to 500 kmph, the simulation is run, for 10 s, to obtain the deflection curves at different modes of the beam (concrete slab), the subgrade, and also at the cement–asphalt mortar joint, for lateral deflection. The Fast Fourier Transform (FFT) is generated for the same, at different modes and at different speeds, for a time duration of 40 s, which in turn gives the magnitude versus frequency graphs.

6 Results and Discussion

Fast Fourier Transform (FFT) for the concrete slab.

6.1 Mode–1

In the above graph, Fig. 5, it can be observed that the maximum value of deflection shoots up at the velocity of 300 kmph, at a frequency of about 2.5 H.z

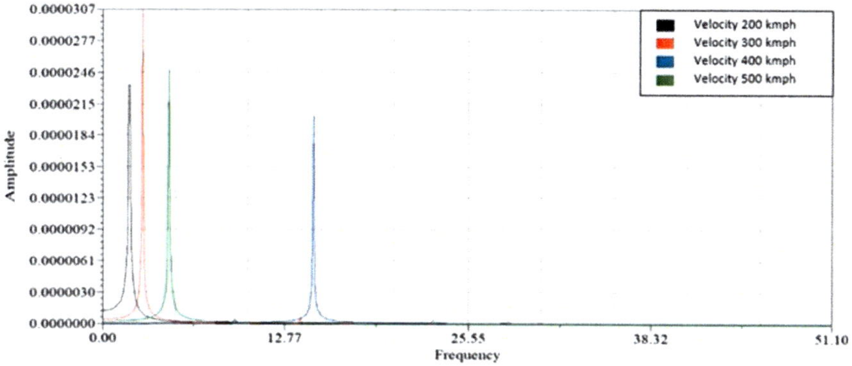

Fig. 5 Deflection (m) versus Frequency (Hz) Fast Fourier Transform curve of Mode 1 of concrete slab at different velocities

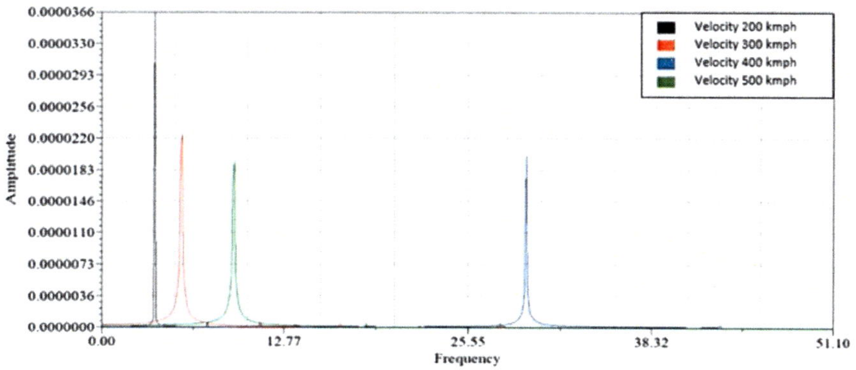

Fig. 6 Deflection (m) versus Frequency (Hz) Fast Fourier Transform curve of Mode 2 of concrete slab at different velocities

6.2 Mode–2

In the above graph, Fig. 6, it can be observed that the maximum value of deflection shoots up at the velocity of 200 kmph, at a frequency of about 3.5 Hz.

6.3 Mode–3

In the above graph, Fig. 7, it can be observed that the maximum value of deflection shoots up at the velocity of 200 kmph, at a frequency of about 5.5 Hz.

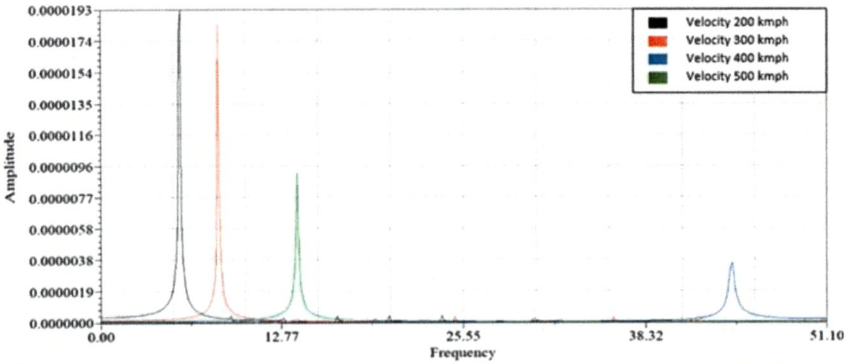

Fig. 7 Deflection (m) versus Frequency (Hz) Fast Fourier Transform curve of Mode 3 of concrete slab at different velocities

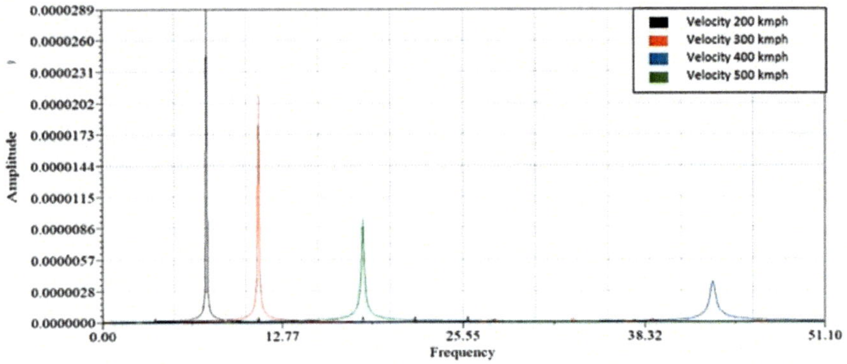

Fig. 8 Deflection (m) versus Frequency (Hz) Fast Fourier Transform curve of Mode 4 of concrete slab at different velocities

6.4 Mode–4

In the above graph, Fig. 8, it can be observed that the maximum value of deflection shoots up at the velocity of 200 kmph, at a frequency of about 7 Hz.

6.5 Mode–5

In the above graph, Fig. 9, it can be observed that the maximum value of deflection shoots up at the velocity of 200 and 300 kmph, at a frequency of about 9 Hz and 13 Hz, respectively.

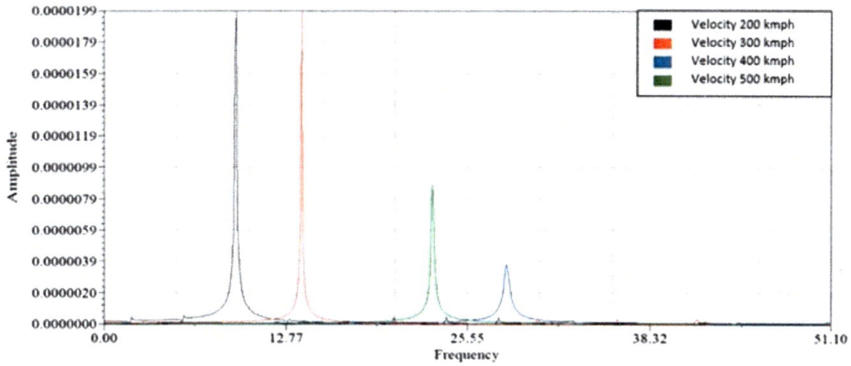

Fig. 9 Deflection (m) versus Frequency (Hz) Fast Fourier Transform curve of Mode 5 of concrete slab at different velocities

7 Conclusions

The Fast Fourier Transform (FFT) of the concrete slab, at different modes, for different velocities, depicts that the excitation of Mode 1 is highest at the speed of 300 kmph, at a value of 0.0307 mm. The excitation of Mode 2 is highest at the speed of 200 kmph, at a value of approximately 0.0366 mm. The excitation of Mode 3 is highest at the speed of 200 kmph, at a value of 0.0193 mm, closely followed by 300 kmph speed, at a value of 0.0183 mm. The excitation of Mode 4 is also highest at the speed of 200 kmph, at a value of 0.0289 mm, and the excitation of Mode 5 is highest at the speed of 200 and 300 kmph, both, at a value of 0.0199 mm. This observation hints that the maximum deflection occurs around the velocity of 200 kmph and in mode 2.

The present study deals with a two-layer problem of beam modelling, in three dimensions, for a concrete slab–subgrade system. In the future, the same methods can be extended for multi-layer bond graph modelling in three dimensions. A more complex system, involving the vehicle as well, on the multi-layered track, can be performed, using bond graph capsules and ports.

References

1. R. Karmakar, A. Mukherjee, Dynamics of electric overhead travelling cranes: a bond graph approach. Mech. Mach. Theory **25**(1), 29–39 (1990)
2. J. Zeng, H. Dai, The dynamic simulation of vehicle-bridge interactions using bond graph technique. Veh. Syst. Dyn. **23**(S1), 591–602 (1994)
3. J.C. Nielsen, A. Igeland, Vertical dynamic interaction between train and track influence of wheel and track imperfections. J. Sound Vib. **187**(5), 825–839 (1995)
4. G. Bureika, R. Subačius, Mathematical model of dynamic interaction between wheel-set and rail track. Transport **17**(2), 46–51 (2002)

5. H. Kumar, B.S.K.C. Sujatha, Lateral dynamic analysis of a typical Indian rail-road vehicle (2005)
6. V. Kumar, V. Rastogi, Lateral dynamic analysis of rail road vehicle through bond graph modeling (2009)
7. V. Kumar, V. Rastogi, P.M. Pathak, Effect of non-linearity on wheel/rail interaction dynamics using bond graph (2018)
8. V. Kumar, Investigation of structural dynamics and ride comfort of rail vehicle system (2018)

Identification of Damping in Sandwich Beam

Harshit Bahri, Kaushalendra Kumar Singh, and Harvendra Singh

1 Introduction

Vibration is the most undesirable factor taken into consideration where there is loading or unloading in structures and relative motion between the members. These uncontrolled vibrations may induce internal stresses which would result in premature failure as discussed by Beards [1]. This reduction in vibrations produced in various structures is in progress by increasing the damping capacity. In the modern aspect, the damping characteristics are enhanced by making lighter machines and also fabricating them in layers instead of monolithic structures. Gould and Mikic [2] observed that the structures which are assembled by welding produce more vibrations in comparison to those fabricated by fastening, as the fasteners such as bolts and rivets are capable of absorbing some part of these vibrations produced.

The modern designs are now focused on the dynamics of mechanical joints as the performance of the structure is strongly influenced by it. Further, the involvement of these joints highly affects the overall behavior of the system, especially the ability to produce damping in the structures. According to Meng and Sun [3] in practical applications, if parameters such as the base layer material and structural loss factor are known, we can compare the behavior of structures by setting the characteristics, i.e., damping material, thickness of layer, and number of layers. However, the evaluation of damping in components subjected to dynamic loading is effectively done by experiment or analysis. The present paper addresses the problem that occurred in the estimation process of damping in beams or beam-like structures with the passive nature of damping. There are various special design techniques that help to improve the capacity of damping in structures, such as structural members build in form of visco-elastic layers, layered sandwich beams, parent material inserted with highly

H. Bahri (✉) · K. K. Singh · H. Singh
Department of Mechanical Engineering, GL Bajaj Institute of Technology and Management, Greater Noida, UP, India

B. Popuri et al. (eds.), *Recent Trends in Engineering Design*, Lecture Notes in Mechanical Engineering, https://doi.org/10.1007/978-981-16-2900-6_2

Fig. 1 Specimen of bolted sandwich beam

elastic material, bolted layered structures jointed, welded and riveted layered structures, etc. may improve the capacity of damping in structures. In a study, Vasina et al. [4] imply about the increase in properties responsible for damping of vibrations in multilayer sandwich structures with the increase in excitation frequency, the thickness of the material and internal mass and decrease in density of the material.

2 Problem Formulation

The specimen for analysis is fabricated from the stock of available mild steel plates and aluminum flat plates. The dimensions of the plates are randomly selected as preferably available in the stock. Further, in order to fabricate the three plates, the temporary fasteners, i.e., bolts of specification M10, are placed in an equi-spaced manner throughout the length of the specimen. The plates are arranged such that the aluminum plate is placed between two mild steel plates which would increase the capacity to damp vibrations of the beam. The distance between all the successive bolts is arranged in such a manner that the regions of pressure distribution near each bolt overlap the other and the uniform pressure distribution is obtained. The bolt diameter as specified above is taken according to the requirement of the beam dimensions. It is necessary to clean the mating surfaces before clamping the specimens with bolts in order to obtain smooth contact between two mating surfaces. The photograph of the involved specimen to be analyzed is shown in Fig. 1.

3 Mathematical Formulation

The problems of vibration can generally be analyzed either by the Euler–Bernoulli beam model or by the Timoshenko beam model. The difference observed between the two models is that in the Timoshenko beam model, the factors taken into consideration are deformation due to shear and effects of rotational inertia, which make it suitable to describe the mechanical behavior of short-length beams, layered sandwich beams, or beams which are subjected to high excitation frequency whereas Euler–Bernoulli

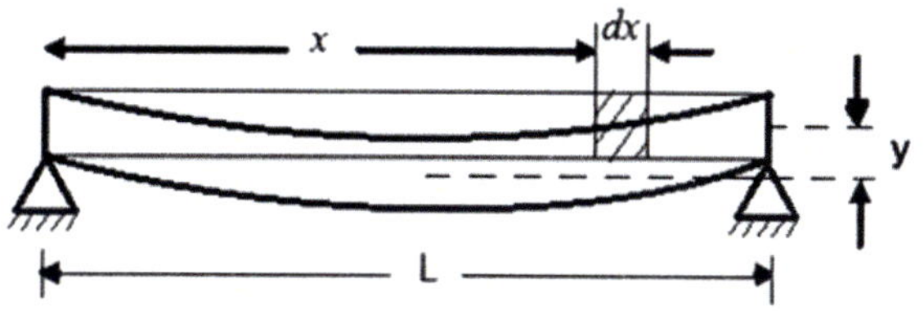

Fig. 2 Beam subjected to free transverse vibration

beam model is a type of classical approach beam model. This model is useful in determining nominal but considerable deflections in the beam as shown in Fig. 2.

The partial differential equations which are expressed in terms of two variables for space function '*x*' and time function '*t*' govern the vibration in beam. Therefore, the differential equation which governs the free transverse vibration of a beam is represented as follows:

$$EI = \frac{\partial^4 y(x,t)}{\partial x^4} + \rho A \frac{\partial^2 y(x,t)}{\partial t^2} = 0 \tag{1}$$

where E: Modulus of elasticity

I: Second moment of area

ρ: Mass density of material

A: Cross-sectional area of beam

The free vibration discussed in Eq. (1) consists of four spatial derivatives, so in order to find a solution, we require four boundary conditions. Also, two initial conditions are required for two time derivatives, out of which, one is for displacement and another is for velocity. *Separation of variables* is the method included in the solution of Eq. (1). The displacement $y(x, t)$ is mathematically expressed as the product of two functions; one is dependent only on space function '*x*' whereas the other is dependent only on time function '*t*'. Thus, the expression of the solution is given as follows:

$$y(x \cdot t) = X(x) * T(t) \tag{2}$$

where $X(x)$ is the space function and $T(t)$ is the time function.

Simplifying the required solution,

We get

$$X(x) = P\ \sin \lambda x + Q\ \cos \lambda x + R\ \sin h\lambda x + S\ \cos h\lambda x \tag{3}$$

Here, by applying the boundary conditions of simply supported beam, we can determine the values of constants P, Q, R, and S.

Now,

$$T(t) = A\ \cos\omega_n t + B\ \sin\omega_n t \tag{4}$$

Thus, the expression for the complete solution for the deflection that occurred in any section of the beam is given as follows:

$$\begin{aligned} y(x,t) = {} & (P\ \sin\lambda x + Q\ \cos\lambda x + R\ \sin h\lambda x + S\ \cos h\lambda x) \\ & * (A\ \cos\omega_n t + B\ \sin\omega_n t) \end{aligned} \tag{5}$$

3.1 Evaluation of Constants P, Q, R, *and* S

By applying the boundary conditions for the simply supported beam,

At $x = 0$,

$$X(0) = 0, \qquad \frac{\partial X(0)}{\partial x} = 0 \tag{6}$$

And at $x = L$,

$$X(L) = 0, \qquad \frac{\partial X(L)}{\partial x} = 0 \tag{7}$$

The required condition of a non-trivial solution for the co-efficient matrix can be fulfilled by this equation which can further be used to evaluate the frequency of vibration. The space function in Eq. (3) can be further expressed in terms of four mathematical equations. The values of Q, R, and S are dependent parameters, whereas P does not depends on any variable and may have any independent value. Taking $P = 1$, we obtain the values of Q, R, and S as follows:

$$P = 1 \tag{8}$$

$$Q = -\frac{\{(\sin\lambda L - \sin h\lambda L)\sin h\lambda L + (\cos\lambda L - \cos h\lambda L)\cos h\lambda L\}}{(\cos\lambda L \sin h\lambda L - \sin\lambda L \cos h\lambda L)} \tag{9}$$

$$R = -1 \tag{10}$$

$$S = -\frac{\{(\sin\lambda L - \sin h\lambda L)\sin h\lambda L + (\cos\lambda L - \cos h\lambda L)\cos h\lambda L\}}{(\cos\lambda L \sin h\lambda L - \sin\lambda L \cos h\lambda L)} \tag{11}$$

Now space function as discussed in Eq. (3) is expressed as follows:

$$X(x) = \sin\lambda x - \left[\frac{\{(\sin\lambda L - \sin h\lambda L)\sin h\lambda L + (\cos\lambda L - \cos h\lambda L)\cos h\lambda L\}}{(\cos\lambda L \sin h\lambda L - \sin\lambda L \cos h\lambda L)}\right]\cos h\lambda x - \sin h\lambda x + \left[\frac{\{(\sin\lambda L - \sin h\lambda L)\sin h\lambda L + (\cos\lambda L - \cos h\lambda L)\cos h\lambda L\}}{(\cos\lambda L \sin h\lambda L - \sin\lambda L \cos h\lambda L)}\right]\cos h\lambda x \quad (12)$$

The various mode shapes of vibration can be given by Eq. (12). In order to get the solution for the four spatial derivatives of free vibration in Eq. (12), we require four boundary conditions. Also, two initial conditions, i.e., one for the velocity and another for displacement are required for the determination of two time derivatives.

3.2 *Evaluation of Constants* A *and* B

Similarly, we get

$$A = \frac{y\left(\frac{L}{2}, 0\right)}{x\left(\frac{1}{2}\right)}, \qquad B = 0 \quad (13)$$

Therefore, substituting all the values in Eq. (5), we get

$$y(x,t) = \left[\sin\lambda x - \left[\frac{\{(\sin\lambda L - \sin h\lambda L)\sin h\lambda L + (\cos\lambda L - \cos h\lambda L)\cos h\lambda L\}}{(\cos\lambda L \sin h\lambda L - \sin\lambda L \cos h\lambda L)}\right]\cos\lambda x - \sin h\lambda x + \left[\frac{\{(\sin\lambda L - \sin h\lambda L)\sin h\lambda L + (\cos\lambda L - \cos h\lambda L)\cos h\lambda L\}}{(\cos\lambda L \sin h\lambda L - \sin\lambda L \cos h\lambda L)}\right]\cos h\lambda x\right] * \left[\left\{\frac{y\left(\frac{L}{2}, 0\right)}{x\left(\frac{L}{2}\right)}\right\}\cos\omega_n t\right] \quad (14)$$

4 Experimental Setup

The experimental setup consists of a beam specimen made up of mild steel layered plates sandwiched with damping element as aluminum and FRF analyzer engaged with a number of supporting instruments. Basically, the setup requires an accelerometer that senses the vibrations produced in the specimen and transmits the signals to the analyzer and also an impact hammer with an inbuilt accelerometer which measures the intensity of vibrations with which the beam is struck. Thus, the analyzed signals are converted into frequency response curves which represent the continuous variations in the frequency of vibrations.

5 Results

In theoretical analysis, the classical approach is preferred to measure the damping of the layered simply supported beam fastened with bolts in a sandwich structure having mild steel plates on top and bottom with an aluminum plate between them. The analysis performed includes the Euler–Bernoulli beam theory summed with some assumptions as it is suitable for small deflections. The strain energy and kinetic energy produced during bending of the beam are used to evaluate stiffness and mass matrices and are further involved in the determination of the natural frequency of vibration and mode shapes by the method of nodal analysis. Subsequently, we have developed the necessary formulations for the logarithmic decrement of both in the case of multi-layered jointed beams.

In order to authenticate the theoretical values determined from the classical approach, they are compared with the experimental values. The observations conclude the comparison between the experimental results and the corresponding from the classical approach, where the maximum variation is about 3.15%. The FRF curves thus plotted are represented in Figs. 3 and 4, respectively.

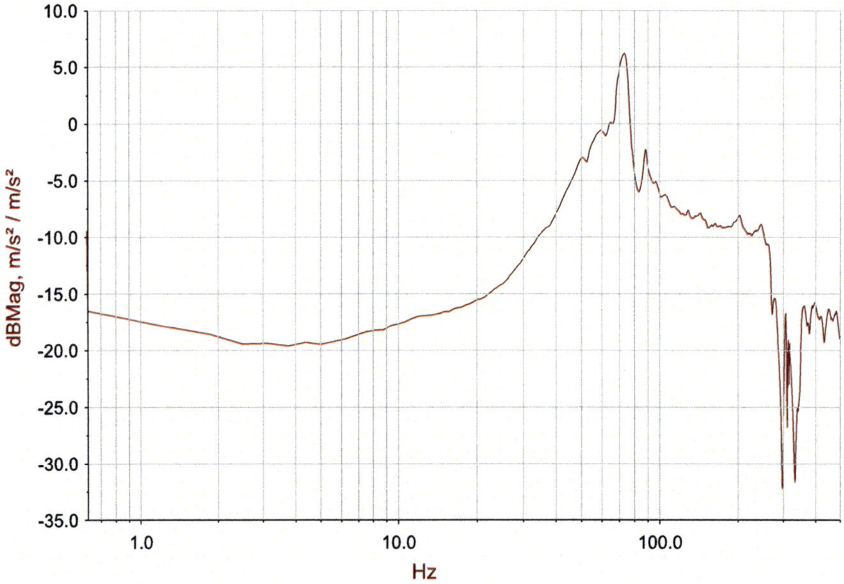

Fig. 3 Amplitude versus frequency curve for the given damped specimen

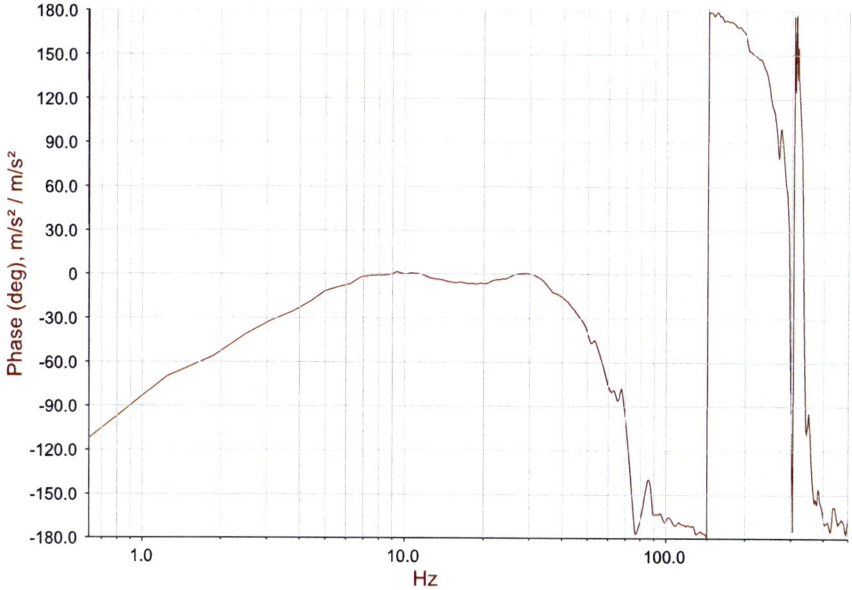

Fig. 4 Phase lag versus frequency curve for the given damped specimen

6 Conclusion

The joints or fasteners used in the fabricated structures are the most important factors responsible for the favorable damping of the structures which are dynamically loaded. The micro-slip occurred and friction developed at the contact between layers are the factors accountable for the energy dissipation taking place. The energy approach because of friction and the dynamic slip is considered for the study of damping in jointed bolted structures. It is found that the appropriate selection of some parameters can enhance the damping of vibrations in layered and jointed structures. These parameters are: (a) torque of tightening each bolt, (b) excitation amplitude, (c) length of given specimen, (d) natural frequency of vibration, (e) thickness ratio, (f) thickness of specimen of beam, etc. Specifically, the thickness of the intermediate damping layer plays a vital role in damping the vibrations produced on loading the beam. The present investigation focuses on a very minute variation in the corresponding values by experimental results when compared with the classical results. The practical approach toward damping in beams appreciates the use of layered structure with structural joints over single-layer solid beams when it is subjected to dynamic loadings. This authenticates the developed theory and the techniques used to determine the logarithmic decrement in layered structures jointed with bolts.

References

1. C.F. Beards, The damping of structural vibration by controlled interface slip in joints. ASME. J. Vibr. Acoust. Stress Reliab. Des. **105**(3), 369–373 (1983)
2. H.H. Gould, B.B. Mikic, Areas of contact and pressure distribution in bolted joints. Trans. ASME. J. Eng. Indus. **94**(3), 864–870 (1972)
3. J. Meng, D. Sun, Research on the multilayer free damping structure design. Shock Vibr. (2018)
4. F. Vasina, L. Hruzik, A. Burecek, Study of factors affecting vibration damping properties of multilayer composite structures. Manuf. Technol. **20**(1) (2020)

Effect of Groove Dimensions on Pressure Profile of Twin Axial Groove Hybrid Journal Bearing

Anamika Yadav and Pooja Pathak

Nomenclature

Λ	Aspect ratio (L/D)
C/R	Clearance ratio
D	Diameter of bearing
H	Film thickness between clearance space
$\bar{X}_j$, $\bar{Z}_j$	Journal center velocity
A	Length of groove
L	Length of bearing
α_g	Location of groove
μ	Lubricant Viscosity
$\bar{K}_\alpha$	Non dimensionless turbulence coefficients in X-direction
$\bar{K}_\beta$	Non dimensionless turbulence coefficients in Y-direction
Ω	Rotational velocity of journal
S_n	Sommerfeld number
P_s	Supply pressure
W	Width of groove

1 Introduction

The foremost and main use of hybrid journal's bearing is their significant applications in turbo-machineries that are generally treated with higher loads and high speeds. Lubricant is generally provided at a pre-specified pressure arising out of the supply

A. Yadav · P. Pathak (✉)
Department of Mathematics, GLA University, Mathura 201406, UP, India
e-mail: pooja.pathak@gla.ac.in

B. Popuri et al. (eds.), *Recent Trends in Engineering Design*, Lecture Notes in Mechanical Engineering, https://doi.org/10.1007/978-981-16-2900-6_3

to the journal bearing's inlet via its grooves which are generally one or two. The geometry of grooves, i.e., its dimensions and quantity of groove, is directly related to the performance of the bearing or we can say that these above-mentioned factors provide an impact on its (bearing's) performance. Such grooves are generally manufactured into the internal surface of the bearings. The groove shape or geometry such as circumferential, spiral, or axial is chosen on the basis of its application. These grooves are generally perpendicular (90°) to the load line and it is in the upstream position of the minimum thickness of the film. The groove location of the bearing strongly affects the stability of turbo machinery system.

Bearing grooves alter the locus of the journal since it expands the angle of attitude and also modifies the cavitated boundary at a lower value of eccentricity and higher speed [1]. The loading direction's effect on double axes groove bearing performance is examined with the help of FEM (finite element method) and the inference made is that the carrying capacity of the load is optimal when the grooves are perpendicular to the load line [2]. An independent measurement of the flow rate for each groove on double axes journal was also incorporated [3].

It is concluded that a single groove present alongside load line on the bearing will perform better. These twin grooves bearing in journal are used dominantly in the condition when the shaft rotates in anti-clockwise as well as in clockwise direction [4]. While carefully analyzing the distribution of pressure for journal bearing having dissimilar configuration or geometry of groove in the hydrodynamic state, it was found that it is better using the circumferential groove located at bearing center, having a hole located for supplying the oil opposite to that of the load carrying zone [5]. An investigative study through some experiments has been carried out, with the aim to investigate the effect of the supply temperature, pressure, and load applied on hydrodynamic bearing's performance having double axes grooves which are perpendicular, i.e., located at ±90° to the load's line, and it was concluded that increasing the inlet pressure, i.e., feed pressure will give rise to increase in the flowing rate (at outlet) of oil. Also, there is a slight fall in the oil's temperature and some rise in its attitude angle, and the oil film thickness is minimum [6, 7]. Influence of the location of the groove on the characteristics in dynamic conditions of journal bearing having multiple grooves journal bearing lubricated through the water was examined, also the stability behavior of journal bearing lubricated through water with multi grooves is studied theoretically along with the calculated stiffness and coefficient of damping for several bearing numbers [8]. A method with the help of using dynamic behavior (characteristics) like the coefficient of stiffness has been taken for determining the hydrodynamic bearing journal stability, i.e., the synchronous' whirl was introduced and found that journal moving at speed of 800 rpm up to 1666 rpm with a load of 150 N remained steady (stable) [9]. Under heavy loading conditions, the groove bearing having double grooves' geometry or configuration could deteriorate the performance of bearing after its comparison was done with a single grooves' geometry configuration as in double groove, there is a non-uniform flow of lubricants in each groove; this finding is made by a comparison done through an experiment comprising a performance-based comparison of a single and twin groove axes bearing [10]. The location of grooves has been set up by considering different arrangements for two-groove journal bearing so that optimum performance could be found that depends upon no dimension load maximization, coefficient of flowing, mass' parameter, and friction variable minimization with the help of developing its genetic algorithm [11]. The turbulence impact on its dynamic working for accelerated or decelerated bearing in hydrodynamic conditions and the analysis of stability for journal bearing of several regimes of flowing were discussed in this work, and the effect of the location of the groove corresponding to the line of load and groove dimensions' influence on its performance was examined. Analysis in the case of a nonlinear system is carried out by its journal's center path (trajectories), by which the linear analysis was verified [12, 13]. The thermo-hydrodynamic analysis has been carried out for two groove and multi-lobe bearing in the clearence space of bearing and shaft [14]. The CFD analysis was done for the thermal characteristics of hydrodynamic bearings having double axes of grooves; this study results that the model proposed can predict the temperature behavior (profile) accurately of the journal bearing, mainly in that portion of the bearing that is inactive and also in groove vicinity [15].

For higher speed and load, hybrid bearing non recessed with grooves is used often. The supply of fluid is maintained by the hole that is located in low-pressure film region. The three main grooves are circumferential, axial, and spiral. The supply pressure is also defined by the grooves. Two-axis grooves are used when there is two side rotation of journal [16]. Journal bearing has very wide applications in turbomachineries such as steam turbines and compressors. The fluid that is flowing in the empty space in the hydrodynamic bearing is not able to remain in the laminar condition because the rotor is running at a very high speed and velocity, and acceleration/deceleration is changing with respect to time. For the equation of trajectories in nonlinear and linear paths, the Runga–Kutta numerical method is used. The effect of grooves geometry, location, and dimensions on static and dynamic characteristics is studied in this paper. Some common equations that are used in the analysis are also discussed. Separation of surfaces that are in relative motion is required before the motion starts; this is the limitation of such bearing. This separation is done by a hydrostatic system that applies a high-pressure fluid in the clearance space.

1.1 Theoretical Analysis Model

The performance analysis of the journal bearing having two grooves has been described with the help of the developed theoretical model. The basic geometry of the bearing is shown in Fig. 1. There are mainly two parameters, i.e., dynamic and static conditions of working on which its operational performance could be divided. The load carrying capacity (reaction of fluid film) comes under the static performance while the dynamic performance consists of the damping and stiffness coefficient of the oil (fluid) film. Eccentricity ratio often represents the measure of the load carrying capacity of the bearing. It depicts the unwrapped geometry of bearings in 2-D and 3-D, respectively, by taking into consideration that the thickness of fluid film is very

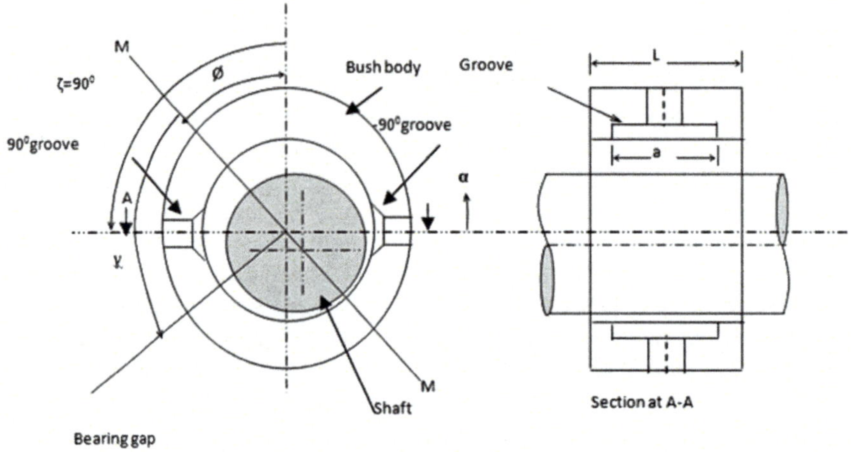

Fig. 1 Bearing geometry [12]

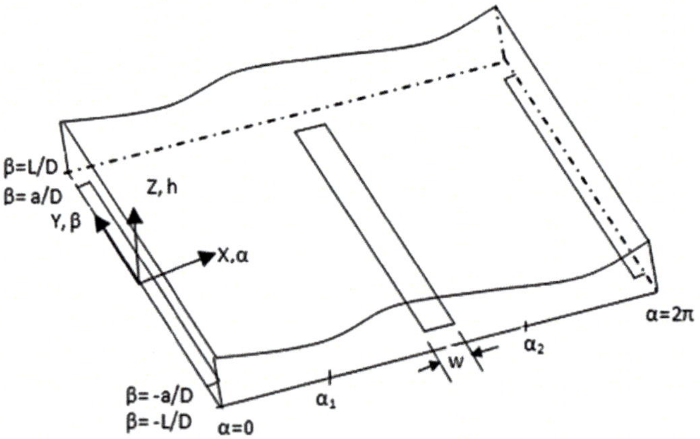

Fig. 2 Development of fluid film between bearing and journal surface (3-D) view [12]

thin. Refer to Fig. 2, in this α_1 and α_2 represent the extent of the degree of the pressure arising in the fluid film. α_{1g} and α_{2g} denote the center line location of the grooves.

1.1.1 Flow Field Equation

The Reynolds equation for non-dimensional case that controls the flow of oil/liquid in journal hydrodynamic bearing's vacant space (clearance space) by using the Constantinescu theory of linearized turbulence (1967) is illustrated in Eq. (1)

$$\frac{\partial}{\partial\alpha}\left[\frac{\bar{h}^3}{\bar{\mu}\bar{K}_\alpha}\frac{\partial\bar{p}}{\partial\alpha}\right]+\frac{\partial}{\partial\beta}\left[\frac{\bar{h}^3}{\bar{\mu}\bar{K}_\beta}\frac{\partial\bar{p}}{\partial\beta}\right]$$
$$=\frac{1}{2}\bar{\Omega}\left(\bar{X}_j\sin\alpha-\bar{Z}_j\cos\alpha\right)-\bar{X}_j\cos\alpha-\bar{Z}_j\sin\alpha \tag{1}$$

The non-dimensional fluid-film thickness $\bar{h}$ for parallel axes case is given by

$$\bar{h}=1-\bar{X}_J\cos\alpha-\bar{Z}_J\sin\alpha \tag{2}$$

1.1.2 Short Bearing Approximation

Supposing that the size of the bearing is taken to be infinitely short, so that the gradient of pressure in axial is much larger than that in circumferential direction as mentioned in the following equation provided:

$$\frac{\partial p}{\partial\alpha}<<\frac{\partial p}{\partial\beta}$$

Then the Reynold equation shortens as Eq. (3) [12]

$$\frac{\partial}{\partial\beta}\left[\frac{\bar{h}^3}{\bar{\mu}\bar{K}_\beta}\frac{\partial\bar{p}}{\partial\beta}\right]=f(\alpha) \tag{3}$$

where

$$F(\alpha)=\frac{1}{2}\left(\bar{X}_j\sin\alpha-\bar{Z}_j\cos\alpha\right)-\bar{X}_j\cos\alpha-\bar{Z}_j\sin\alpha$$

Equation (3) is solved using new boundary conditions as given in Sect. 1.1.3.

1.1.3 Boundary Condition

The applicable conditions of boundary as per the given problem are as given in Eq. (4)

(i) $$\bar{p}=0\ \text{at}\ \beta=\pm\frac{L}{D}=\pm\lambda \tag{4}$$

After doing the integration of Eq. (3) with β and applying the conditions of boundary, Eq. (4) that indicates the distribution of pressure about journal is specified as follows:

$$\bar{p}(\alpha,\beta)=\left(\frac{\bar{\mu}\bar{K}_\beta}{\bar{h}^3}\right)\left(\frac{1}{2}f(\alpha)\left(\beta^2-\lambda^2\right)\right) \tag{5}$$

After obtaining pressure distribution, all negative pressure is made zero. Further, the following conditions were also applied.

(1) $$\bar{p} = \bar{p}_s \quad \text{at} \quad \left[\left(-\frac{a}{D} < \beta < \frac{a}{D}\right) \text{ and } \left(\alpha_{1g} - \frac{W}{D}\right) < \alpha < \left(\alpha_{1g} + \frac{W}{D}\right)\right] \tag{6a}$$

(2) $$\bar{p} = \bar{p}_s \quad \text{at} \quad \left[\left(-\frac{a}{D} < \beta < \frac{a}{D}\right) \text{ and } \left(\alpha_{2g} - \frac{W}{D}\right) < \alpha < \left(\alpha_{2g} + \frac{W}{D}\right)\right] \tag{6b}$$

where α_{1g} and α_{2g} are the center lines of the angular position of grooves in radian.

1.1.4 Load Carrying Capacity

On integrating pressure w.r.t zone for positive pressure, the capacity of carrying the load of the journal bearing could be found. The capacity of carrying of load in the circumferential direction is mentioned as follows:

$$\bar{F}_X = -\int_{-\lambda}^{\lambda}\int_{0}^{2\pi} \bar{p} \cdot \cos\alpha d\alpha \cdot \mathrm{d}\beta \tag{7}$$

and similarly, the load carrying capacity in the radial direction (Z) is written as follows:

$$\bar{F}_Z = -\int_{-\lambda}^{\lambda}\int_{0}^{2\pi} \bar{p} \cdot \sin\alpha \mathrm{d}\alpha \cdot d\beta \tag{8}$$

On substituting the value of $\bar{p}$ as found from Eq. (5) in Eq. (7) The $\bar{F}_X$, i.e., the carrying capacity of load is mentioned as follows:

$$\bar{F}_X = -\int_{0}^{2\pi}\left(\frac{\bar{\mu}\bar{K}_{\bar{\beta}}}{\bar{h}^3}\right)\frac{f(a)}{2}\cos\alpha\left(\int_{-\lambda}^{\lambda}(\beta^2 - \lambda^2)d\beta\right)\mathrm{d}\alpha$$

Let

$$f_1(\alpha) = \frac{\bar{\mu}\bar{K}_{\beta}}{\bar{h}^3}\left(\frac{f(a)}{2}\right)$$

then

$$\bar{F}_X = -\int_0^{2\pi} f_1(\alpha)\cos\alpha\left(\int_{-\lambda}^{\lambda}(\beta^2 - \lambda^2)d\beta\right)d\alpha \tag{9}$$

Similarly, for Z-direction, the load carrying capacity in non-dimensional form is given by

$$\bar{F}_Z = -\int_0^{2\pi} f_1(\alpha)\sin\alpha\left(\int_{-\lambda}^{\lambda}(\beta^2 - \lambda^2)d\beta\right)d\alpha \tag{10}$$

Equations (9) and (10) are solved numerically using the Gauss–Legendre numerical integration over the positive pressure zone.

1.2 Computational Procedure

The Reynolds equation for lubricant flow in the case of the journal bearing is altered for studying the kind of flow like laminar flow, transition, or the turbulent one, by considering $\bar{K}_X$ and $\bar{K}_Y$ as its turbulence coefficients. In order to solve the Reynolds equation, approx. of short bearing was used. By considering this approximation of short bearing and the Reynolds equation have been integrated twice, by inserting the new boundary conditions in the corresponding limits and the pressure expression in closed form was obtained. Pressure zone (positive) has been set up by eliminating the pressure below atmospheric (sub-ambient), and 100 kPa supply pressure (p_s) is taken at the entire groove area. A computer program with modified boundary conditions was developed.

1.3 Results and Discussion

The algorithm of solution and the analysis were employed for computing the profile of the pressure, the reaction of oil or fluid film or loading capacity, stiffness of the fluid film, and coefficient of damping. The taken values of bearing Eccentricity or Aspect ratio, i.e., (L/D) ratio, are 0.5 and 0.25, respectively, for conducting these studies. Taken the assumption that the axis of bearing and axis of journal are parallel to each other and the nominal clearance to the radius of journal ratio, i.e., C/R = 0.001.

1.3.1 Validation of Results

To establish the validity of the analysis, solution algorithms, and the computer program, the circumferential pressure at the mid-plane of the bearing obtained from the present short bearing approximation was compared with results of Kostrzewsky

Fig. 3 Comparison of pressure profile at mid-plane of the bearing with present analysis

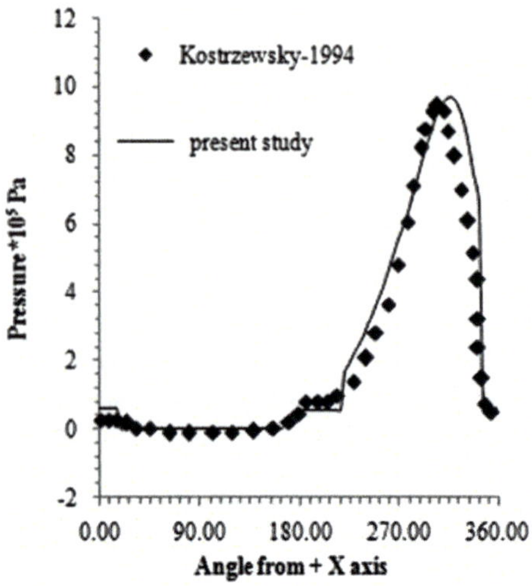

et al. [17, 18] as shown in Fig. 3. The comparison of circumferential pressure with angular location from +X-axis is shown for L/D ratio of 0.50, C/R of 0.00294, and speed 1125 rpm.

The comparison of circumferential pressure with angular location from −X-axis is shown for L/D ratio of 0.50, C/R of 0.001, viscosity of 0.0111 Pa · s, and speeds 2000 and 4000 rpm.

Results could be correlated appropriately if there are some changes because of variation in the chosen methodology in the solutions. After the procedure of solution has been verified or validated, results that have come out for distinct dimensions of the grooves, locations, and condition of flowing for super laminar are discussed in the later sections of this paper.

Post validating the procedure of solution, the following sections/paragraphs result the study for the influence of dimensions of grooves, their possible locations, and the conditions for super laminar flowing of liquid.

1.3.2 Influence of the Groove Length Ratio (a/L) on Pressure Profile

Initially, the pressure profile is obtained by varying the length of the grooves while width and angular position of grooves were kept fixed as $w/D = 0.05$ and the angle of the first groove at $\alpha_{1g} = 0°$ and second groove at $\alpha_{2g} = 180°$ (Fig. 2). The effect of groove length ratio (a/L) on pressure profile is studied in axial and circumferential directions. These variations are shown in Fig. 4, 5, and 6 for the values of $a/L = 0.8$, 0.6, and 0.4, respectively.

It was observed that the total pressure decreases with the increase in groove length ratio as groove reduces the extent of pressure field. The location of the maximum pressure is in the middle of the axial direction and at 295.2° in the circumferential

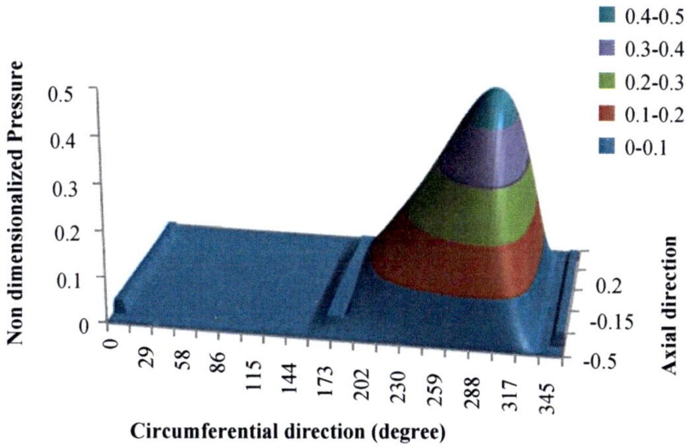

Fig. 4 Variation of pressure profile for a/L = 0.8

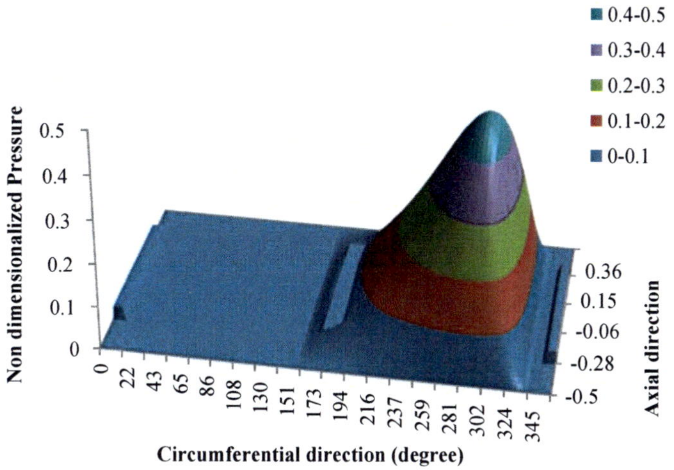

Fig. 5 Variation of pressure profile for a/L = 0.6

direction measured from the line M-M, Fig. 1. The value of the maximum non-dimensionalized pressure is equal to 0.45386 for a/L = 0.2. The location of maximum non-dimensional pressure in the circumferential direction shifts with decrease in a/L ratio. It was at angular position of 293° for a/L = 0.8 whereas for a/L = 0.6, it is 294.5°, for a/L = 0.4, it is 295°, for a/L = 0.2, it is 295.2°, and for a/L = 0.1, it is 296°. The location of the maximum pressure is in the middle of the axial direction and at 295.2° in the circumferential direction measured from the line M-M, Fig. 1. The value of the maximum non-dimensionalized pressure is equal to 0.45386 for a/L = 0.2. The location of the maximum non-dimensional pressure in the circumferential direction shifts with a decrease in a/L ratio. It was at an angular position of 293° for

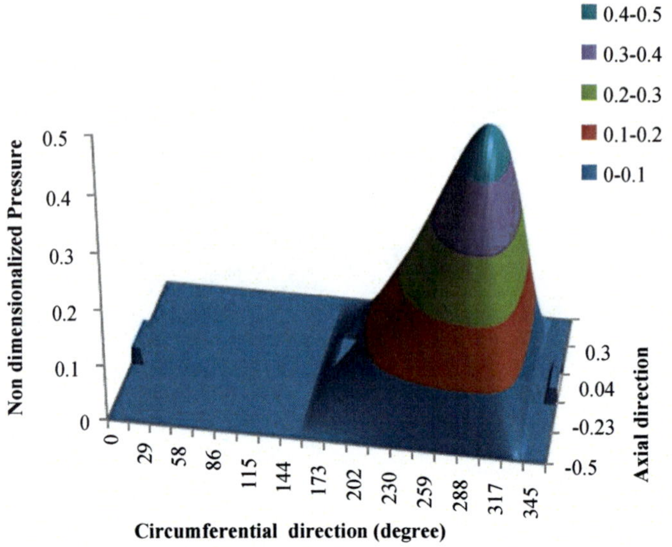

Fig. 6 Variation of pressure profile for a/L = 0.2

a/L = 0.8 whereas for a/L = 0.6, it is 294.5°, for a/L = 0.4, it is 295°, for a/L = 0.2 it is 295.2°, and for a/L = 0.1, it is 296°.

1.3.3 Influence of the Groove Width Ratio (W/D) on Pressure Profile

The pressure profile is obtained by varying the width of the grooves while fixing the groove length ratio and angular position of grooves. The groove ratio was taken as a/L = 0.1. Location of groove was fixed by taking the angle of the first groove at $\alpha_{1g} = 0°$ and second groove at $\alpha_{2g} = 180°$ (Fig. 2). The effect of groove width ratio (w/D) on pressure profile is studied in the axial and circumferential directions. Figures 7 and 8 show the hydrodynamic pressure profile in the entire plane of the bearing for groove width ratio w/D = 0.1 and 0.2, respectively.

It was observed that the maximum pressure decreases with the increase in groove width ratio due to the spread of the groove in the circumferential direction. When w/D ratio increases, the position of the maximum pressure shifts toward the bearing ends and if the w/D ratio is equal to one, the bearing will start working as the bearing with a circumferential groove has two peak pressures along the axial direction.

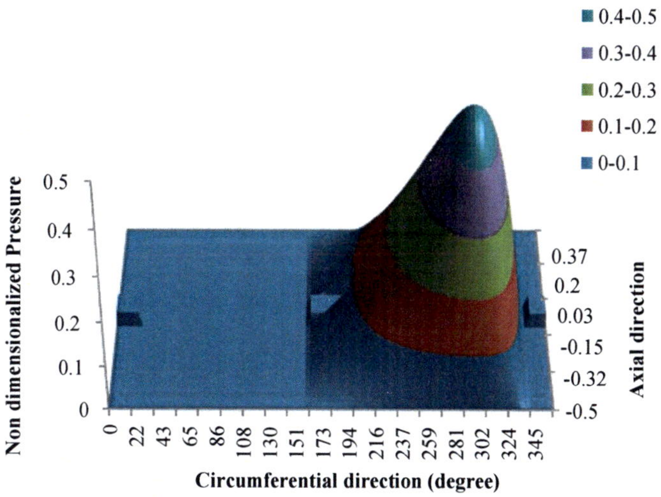

Fig. 7 Variation of pressure profile for groove width ratio, w/D = 0.1

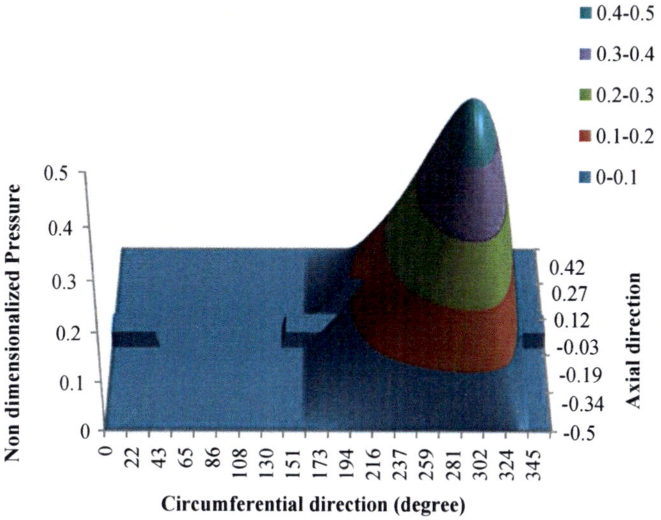

Fig. 8 Variation of pressure profile for groove width ratio, w/D = 0.2

1.4 Conclusion

A detailed study along with discussions on the journal bearings has been briefed in this paper. A parametric performance study with the help of algorithms and numerical techniques due to the supply of the lubricant through grooves has also been discussed in this paper. Result clearly illustrates that the groove length ratio and width ratio should be 0.8 and 0.1 in order to have enhanced load bearing capacity. Optimal groove length and width ratio that came out to be 0.8 and 0.1, respectively, with the help of algorithms, i.e., numerical and programming techniques (computational model) were presented.

References

1. P.G. Morton, J.H. Johnson, M.H. Walton, The influence of grooves in bearings on the stability and response of rotating systems. Tribol. Ser. **11**, 347–354 (1987)
2. Gethin, D.T., El Delhi, M.K.I.: Effect of loading direction on the performance of a twin-axial groove cylindrical-bore bearing. Tribol. Int. **20**(4), 179–185(1987)
3. Basri, H., Neal, P.B., *Oil flow in axial groove journal bearings*. Instintuition of Mechanical Engineering Seminar, Devision of Plain Bearing, London (pp. 11–17) (1990)
4. L. Costa, A.S. Miranda, M. Fillon, J.C.P. Claro, An analysis of the influence of oil supply conditions on the thermohydrodynamic performance of a single-groove journal bearing. Proc. Inst. Mech. Eng. Part J. Eng. Tribol. **217**, 133–144 (2003)
5. C. Desai, D. Patel, Experimental analysis of pressure distribution of hydrodynamic journal bearing: a parametric study. Proc. Int. Conf. Mech. Eng. 1–4 (2005)
6. F.P. Brito, J. Bouyer, M. Fillon, A.S. Miranda, Thermal behaviour and performance characteristics of a twin axial groove journal bearing as a function of applied load and oil supply temperature. J. Tribol. **25**(3), 24–33 (2006)
7. F.P. Brito, A.S. Miranda, J. Bouyer, M. Fillon, Experimental investigation of the influence of supply temperature and supply pressure on the performance of a two-axial groove hydrodynamic journal bearing. J. Tribol. **129**(1), 98–105 (2007)
8. M.V. Kini, R.S. Pai, D.S. Rao, S.B. Satish, R. Pai, Effect of groove location on the dynamic characteristics of multiple axial groove water lubricated journal bearing. World Acad. Sci. Eng. Technol. **3**(12), 1592–1596 (2009)
9. R.R. Navthar, N.V. Halegowda, Stability analysis of hydrodynamic journal bearing using stiffness. Int. J. Eng. Sci. Technol. **2**(2), 87–93 (2010)
10. F.P. Brito, A.S. Miranda, J.C.P. Claro, M. Fillon, Experimental comparison of the performance of a journal bearing with a single and a twin axial groove configuration. Tribiol. Int. **54**, 1–8 (2012)
11. L. Roy, S.K. Kakoty, Optimum groove location of hydrodynamic journal bearing using genetic algorithm. Adv. Tribol. Article ID 58036 (2013). https://doi.org/10.1155/2013/580367
12. V.K. Dwivedi, S. Chand, K.N. Pandey, Effects of turbulence on dynamic performance of accelerated/decelerated hydrodynamic journal bearing system. Int. J. Des. Eng. **5**(3), 256–288 (2014)
13. V.K. Dwivedi, S. Chand, K.N. Pandey, Stability analysis of twin axial groove hybrid journal bearing. J. Appl. Fluid Mech. **9**(6), 2763–2768 (2016)
14. L. Roy, C. Bhagat, Tribology in industry steady state thermo-hydrodynamic analysis of two-axial groove and multilobe hydrodynamic bearings. Tribol. Int. **36**, 475–487 (2015)
15. A.A. Solghar, F.P. Brito, M. Abdolzadeh, A. Farajpour, Numerical study of twin groove journal bearings performance under steady-state condition. Lubr. Sci. **27**, 83–102 (2015)
16. T. Ren, M. Feng, Tribology International Stability analysis of water-lubricated journal bearings for fuel cell vehicle air compressor. Tribiol. Int. **95**, 342–348 (2016)
17. G.J. Kostrzewsky, D.V. Taylor, R.D. Flack, Experimental determination of the dynamic characteristics of a two-axial groove journal bearing. Tribol. Trans. **37**(3), 534–542 (1994)
18. G.J. Kostrzewsky, R.D. Flack, Accuracy evaluation of experimental derived dynamic coefficients of Fluid-Film Bearings, Part I: Development of the Method. Tribol. Trans. **33**(1), 105–114 (1990)

Method of Obtaining Planar State of Stress Using Mohr's Circle—Some Typical Cases

Shailendra Singh Chauhan and Avadhesh Kumar Khare

1 Introduction

The tensile and shear stresses at a point P in any stressed material keep on changing from one plane to another plane passing through the same point P, and it is not possible to keep a record of these stresses along all the infinitely large number of planes. Fortunately, it is only necessary to keep a record of tensile and shear stresses at the point P on a finite number of planes passing through that point.

Analytical and Graphical methods are available in the literature describing the procedure of finding the State of Stress at a point P and obtaining the normal and shear stresses on any plane passing through that point [1–3]. Analytical methods and Mohr's Circle for 3-D have been discussed in detail in [4]. A simple rule-based method for constructing Mohr's Circle for plane stress was explained and the sign conventions for shear stresses were discussed to eliminate anomaly arising out of the explanation given in some textbooks in [5]. The modified form of Mohr's criteria is suggested in [6] to overcome certain difficulties in the Mohr–Coulomb theory for isotropic brittle materials. Mohr's diagram for anisotropic diffusion tensor in MRI was presented in [7]. The concept of Pole Point and its construction is explained in [8] which eliminates remembering that the angle in Mohr's Diagram is twice the angle in the space diagram, which seems confusing to some. The two different sign conventions widely used for shear stresses are discussed in relation to Mohr's Diagram and the method of Transformation of equations and Mohr's Circle is compared in [9]. The three-dimensional Mohr's Strain circle is discussed in [10] which also establishes powerful applications of Mohr's Circles, both stress circles and strain circles. Whereas, analytical procedures are more suitable for 3-D state of

S. S. Chauhan (✉)
GL Bajaj Institute of Technology and Management, Greater Noida, Gautam Buddha Nagar 201310, UP, India

A. K. Khare
Quantum University, Roorkee, India

B. Popuri et al. (eds.), *Recent Trends in Engineering Design*, Lecture Notes in Mechanical Engineering, https://doi.org/10.1007/978-981-16-2900-6_4

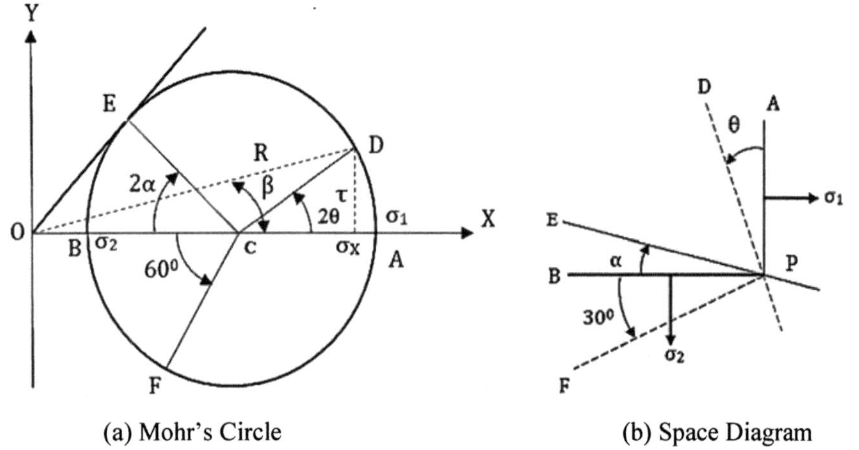

Fig. 1 Space diagram and Mohr's diagram

stress; the graphical method is very convenient and easy for 2-D problems. If Mohr's Circle can be constructed uniquely, then naturally the normal and shear stresses on any plane passing through that point can be obtained.

There is always a one-to-one correspondence between the Space Diagram representing the planes passing through a point P say 'A' and Mohr's Diagram representing the normal and shear stresses on the planes passing through that point. Referring to Fig. 1, the principal planes A and B (in the space diagram) are represented by the radius vectors CA and CB, respectively, in Mohr's Diagram. The angle between any two planes in the space diagram is one half of the angle between the two corresponding radius vectors representing these planes, but in the same direction.

Referring to Fig. 1b, the planes D, E, and F passing through the same point P in the space diagram correspond to the radius vectors CD, CE, and CF, respectively, in Mohr's Diagram Fig. 1a. It may be noted that, for example, F plane is obtained by rotating B plane through 300 in CCW direction in the space diagram Fig. 1b. And therefore, radius CF is obtained by rotating CB through 600 (two times 300) in CCW direction in Mohr's Diagram Fig. 1a. In Mohr's diagram, radius CD is obtained by rotating radius CA through 2θ in CCW direction, and plane D is obtained by rotating plane A through an angle θ (half of 2θ) in CCW direction in the space diagram. The radius CE represents a plane E in the space diagram, where the total stress vector at point P is most inclined to the normal of the plane. Note that the angle BCE in Mohr's diagram is twice the angle between the planes B and E in the space diagram, both measured in the same CW direction. The same is true, as a matter of fact, for any other plane also. This establishes a one-to-one correspondence between the Space Diagram representing the planes at a point P and Mohr's Diagram represented by the radius vectors.

2 Stresses Partially Known on Three Planes

2.1 *Case 1*

Referring to Fig. 2, the stresses are given partially on three planes. At point P, the normal stresses on three planes, namely A, B, and D are known to be +20, +100, and −60 N/mm^{2,} and the shear stresses are not given. The planes A and B are separated by angle 60° CCW, and planes B and D are separated by angle 50° CCW as shown in Fig. 2b. Then Mohr's Circle can be constructed using the steps given below.

(1) Draw Y-axis YY; (2) Set three vertical lines representing +20, +100, and −60 N/mm^2 representing planes A, B, and D; (3) Take arbitrary point K on +20 N/mm^2 line; (4) Note from the space diagram that planes B and , and planes D and B are separated through angles 60° and 50° CCW, respectively; (5) Rotate vertical line through K in the CCW direction by 60° to cut +100 N/mm^2 line at B; (6) Rotate KB by 50° CCW and extend backwards to cut −60 N/mm^2 line at D; (7) Locate the center of Mohr's Circle at C as the point of intersection of perpendicular bisectors of KB and KD; (8) Complete the circle passing through B, K, and D; (9) Draw X-axis NOCM by drawing horizontal through C; (10) Locate A as the mirror image of K on X-axis; (11) Note that angles ACB, BCD, and DCA are 120°, 100°, and 140°, respectively. Also, the angle subtended by any chord of a circle at the center is twice the angle subtended by that chord at the circumference. Since it is possible to construct a unique Mohr's Circle, the State of Stress at the point P is known. Thus,

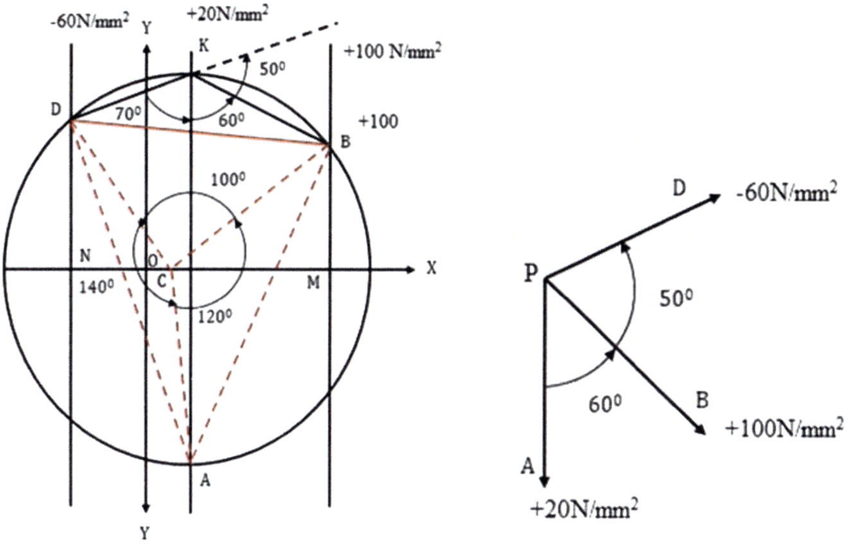

(a) Mohr's Circle (b) Space Diagram

Fig. 2 Space and Mohr's diagrams for Case 1

the state of stress at a point is known if normal stresses on each of the three planes are known and the two angles between three planes are given.

2.2 *Case 2*

Referring to Fig. 3, at a point P in a stressed material, shear stresses are given on three planes A (-40 N/mm^2), B ($+10$ N/mm^2), D (-25 N/mm^2), and normal stress is given on plane A ($+10$ N/mm^2). Shear stress is taken as positive if it results in a CW moment about point P. Plane B is separated through an angle of 60° CCW from plane A. The angle between plane D and plane B is unknown and can be taken as θ CCW. Mohr's Circle can be constructed using the steps given below.

(1) Draw the Coordinate axes OX and OY; (2) Draw lines parallel to X-axis to represent shear stresses (-40 N/mm^2), ($+10$ N/mm^2), and (-25 N/mm^2); (3) Mark the point A having coordinates ($+10$, -40 N/mm^2); (4) Mark point K as the reflection of point A about X-axis; (5) Rotate the line KA about point K by 60° CCW same as the plane A is rotated in space diagram to get plane B; (6) The line KB intersects the $+10$ N/mm^2 horizontal line at point B; (7) Locate the center of Mohr's Circle at C as the point of intersection of the perpendicular bisector of KB and X-axis; (8) Complete the circle passing through B, K, and A and with C as the center; (9) The circle intersects the -25 N/mm^2 line at point D; (10) Note that the angle ACB is 120° CCW, and the angle DCB is 2θ CCW; (11) The construction is validated by knowing that the angle subtended by any chord of a circle at the center is twice the angle subtended by that chord at the circumference.

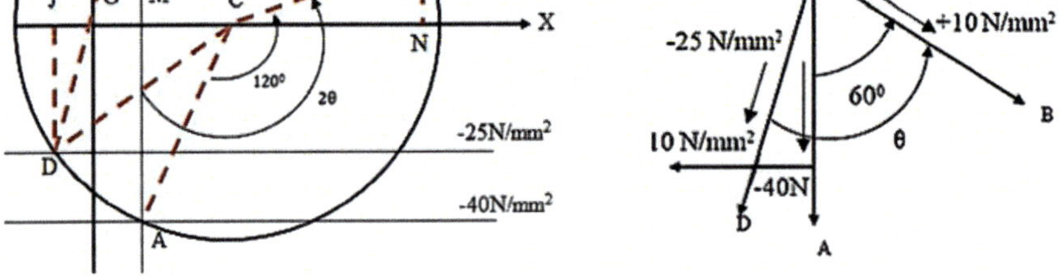

(a) Mohr's Circle (b) Space Diagram

Fig. 3 Space and Mohr's diagrams for Case 2

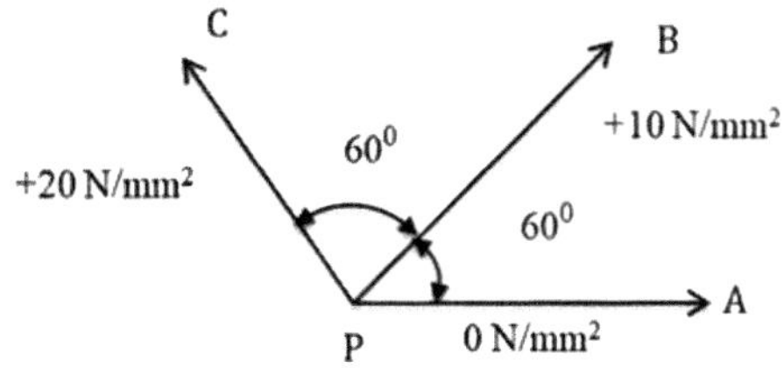

Fig. 4 Space diagram for Case 3

Since it is possible to construct a unique Mohr's Circle, the State of Stress at the point P is known. Thus, the state of stress at a point is known if three shear stresses on three different planes and normal stress on one plane are given, besides one angle between any two planes.

2.3 Case 3

At a point P in the stressed material, if only shear stresses are given on three planes separated by known angles, then it may not always be possible to construct Mohr's Circle. For example, on three planes A, B, and C at a point P, the planes A and B are separated by angle 60° CCW, and planes B and C are separated by the same angle 60° CCW in Fig. 4. If the shear stresses are known to be 0 N/mm^2 on A, +10 N/mm^2 on B, and +20 N/mm^2 on C, then the given data is inconsistent and Mohr's Circle cannot be constructed. It may be noted that plane A is a Principal Plane, in Mohr's diagram, planes B and C will be at angular distances 120° and 240° from plane A, and therefore must have equal and opposite shear stresses. Since the given shear stresses on planes B and C are +10 N/mm^2 and +20 N/mm^2, respectively; therefore, the given data is inconsistent.

3 Conclusion

In 2-D stress system, the Graphical method is very conveniently used to find the state of stress at any point P and calculate the normal and the shear stresses on any given plane at that point. Generally, the stresses are given on two planes that are mutually perpendicular or inclined at an angle. Then the problem of constructing Mohr's diagram is simple.

However, in some situations, stresses are partially known on three planes. It is noted that if three normal stresses are given on three planes separated at known angles, it is possible to draw Mohr's diagram. Also, if three shear stresses are given on three planes with one normal stress known, and only one angle is given, then also it is possible to draw Mohr's diagram. But in some of the cases, such as three shear stresses and two angles are known, the given data may be inconsistent and Mohr's Circle could not be constructed.

References

1. S.P. Timoshenko, D.H. Young, *Elements of Strength of Materials*, 5th edn. (Affiliated East-West Press, New Delhi, 2009), pp. 59–61
2. G.H. Ryder, *Strength of Materials*, 3rd edn. (Macmillan Publishers India Ltd., Chennai, 2009), pp. 42–44
3. A. Pytel, F.L. Singer, *Strength of Materials*, 4th edn. (Addison Wesley, India, 1999), pp. 326–330
4. S.-S. Xu, A.F. Nieto-Samaniego, S.A. Alaniz-Álvarez, 3D Mohr diagram to explain reactivation of pre-existing planes due to changes in applied stresses, in *Rock Stress and Earthquakes*, ed. by Xie (Taylor & Francis Group, London, 2010). ISBN 978-0-415-60165-8
5. S. Krishnamoorthy, A rule based method to construct the Mohr's circle for plane stresses. Glob. J. Eng. Sci. (2019). ISSN: 2641-2039. Accessed 20 Nov 2019
6. A.M. Comanici, Modification of Mohr's criterion in order to consider the effect of intermediate principal stress. Int. J. Plast (2018). https://doi.org/10.1016/j.ijplas.2018.04.010
7. M. Bilgen, I. Elshafiey, P.A. Narayana, Mohr's diagram for anisotropic diffusion tensor in MRI. Magn. Reson. Med. **47**, 823–827 (2002)
8. R. Brannon, Mohr's circle and more circles, in *Welche Umsta_umlaut Finish Citation from Jaeger and Cook Rock Mechanics Book* (Mohr 1900 October 29, 2003)
9. M. Du, G. Chen, K. Pu, Simplification of the parametric transformation equations for analysis of plane stress. Adv. Mater. Res. **591–593**, 2499–2503 (2012)
10. S.H. Treagus, Some applications of the Mohr diagram for three-dimensional strain. J. Struct. Geol. **8**(7), 819–830 (1986)

Computational Modeling of Bubble Formation on Submerged Orifice

Sachin Kumar and Raj Kumar Singh

1 Introduction

Profile shape on bubble formation in the liquid is one of the major questions that arises in many physical and chemical process in different industries which work on it because diffusion of a gas bubble in the liquid is one of the main factors to consider in industries in metal forming or iron rod formation, detailed studies on this topic is a major impact on the high-temperature metal forming processes such as steel component formation, iron rod formation or metal forming process or metal refining process. Removal of gas droplets from liquid metals and non-metallic gas is trapped in the porous plug, slag foaming process in bath inhaling are all surface process at the triple-layer interfaces. Hence, we say that the bubble size is the main control factor in the various process for the making of steel material or alloy formation in many industries in simple words we say that the bubble size and the wetting conditions are the main factors to minimize in the industries process. In industries, the bubble evolution is a very important topic to research for engineer and scientist and it has been studied last 60 years [1, 2], which involved detailed study of every aspect of bubble and study has been done. "Different investigated models [3, 4], from simple 'one-stage has more authentic two-stage models, were to give the bubble evolution process by calculating the bubble of rising force like buoyant force, pressure force and momentum of gas and restraint number of forces will act on a bubble, e.g., drag, surface tension, and inertia forces [5, 4, 6]". The buoyant force is made equilibrium by the surface tension at a very small gas flow rate; this evolution is known as static bubble state and bubble cubic measure changes with the nozzle rim and the surface tension and gas flow rate does not depend. The analytic model forecasts the volume of a bubble at this state [6]. Surface tension is near to zero apart from the initial stage of bubble evolution and the inertia force dominants over the surface tension force. Initial steps of study of Davidson and Schüler [2, 5] movement of the bubble in

S. Kumar (✉) · R. K. Singh
Department of Mechanical Engineering, Delhi Technological University, Delhi 110042, India

B. Popuri et al. (eds.), *Recent Trends in Engineering Design*, Lecture Notes in Mechanical Engineering, https://doi.org/10.1007/978-981-16-2900-6_5

viscous and non-viscous liquids, the advance study gave the using of two-stage and non-spherical models [7]. In the number of studies the gas flow through the nozzle, i.e., the flow rate of gas is constant or pressure is a constant state, was effectively checked. Boundary-integral methods were used by Higuera [8]. At the high flow of air, the bubble evolution repetition remains to keep fixed and the bubble cubic measure changes with the airflow rate and is independent with the surface tension force. The periphery of the bubble increases with the flow rate of air but these studies have taken the assumption of repetition of bubble and do not depend on the already formed bubble on the evolution of the bubble at the nozzle.

In the recent study, numerical simulation has used a couple of the volume of fluid and level-set methods where using Navier–Stokes equations in the symmetric mathematics was solved in the computational geometry. We focus on the flow rate of gas and nozzle radius and effects of the surface tension, the density of the liquid, and the resistance force acting between the layers of fluid. If the flow rate of gas increased single bubble formation detaches earlier as compared to double periodic formation. The experiment was done by Kyriakides et al. [9], Zhang and Shoji [10], and Tufaile and Sartorelli [11]. As we see the first bubbles in the pairing of the bubble are bigger than the second bubble, and provide all relevant data under which the formation state changes. We take the reference as air and water (20 °C) and fluid properties changed individually to study the impact on the bubble formation and not give any material properties that were provided in literature as discussed by Ponter and Surati [12], which does not compare the result between different studies so it is very difficult to analyze the orifice material. Almost all models are based on some hypothesis that the bubble-based corresponds to the nozzle rim during evolution, which can be a restraint in the nozzle surface material employed. Based on the previous study, the recent study gives the result of bubble evolution at submerged nozzle under given boundary conditions and the present simulation result approach can give a better description of the bubble evolution process. In previous studies, we concentrate on one of the impacts of the fluid properties on the bubble evolution process in which the flow rate is constant [4, 6, 13], the time work takes the properties in combination with bubble interaction and the wetting condition of the nozzle plate. Buwa et al. [14] research the bubble motion in a nozzle of 2–10 mm, associated with the computational work. In their work, the flow rate of air and bubble pairing and collision of the bubble at constant nozzle diameter were studied. Gerlach et al. [15] found the same bubble behavior and observed the single bubble evolution and double bubble evolution process in their numerical approach. Hong et al. (1996) and Yang et al. (2007a, b) gave the study of a time-dependent variable around the bubble periphery and the interaction on the gas liquid, and solid-phase when the bubble motion during the bubble particle collision. Li et al. (2000) gave a computational study about the bubble evolution in a vertical cylinder under an immense pressure state. The study gave the result that the immense pressure gives a smaller bubble volume in a bubble cylinder. Ansys Fluent simulations have been studied various parameters of the bubble evolution at adiabatic condition (Quan and Hua 2008; Das and Das 2009; Chakraborty et al. 2009; Ohta et al. 2011; Chakraborty et al. 2011) and give the result which is very useful. Dynamics of fluid and bubble evolution have been done in the latest years, in which a Multiphase model volume of fluid takes the intricate gas–liquid transition line in the domain. Zhou et al. (2011) taken the multiphase model which used the VOF method to simulate the processes of air bubble initialization, growth, swelling of the bubble, and disengagement. This work mainly concentrated on the uses of wetting of the wall where the bubble gets initialized. The computational results showed that the horizontal base wall provided the release of bubble emerged. The software simulation has given that the horizontal base wall provided the discharge of bubbles while the air coverage ratio is getting increased and the dimensionless flow resistance (2013) used a couple of level-set methods and the VOF method to simulate the bubble size increases in static liquids and the eruption occurred at the atmospheric condition important parameter study give the clear understanding of bubble mechanism. This matter is discussed in a large scale due to the metal forming process which includes the trapping of air process in porous plugs, nozzles, and orifices.

The most recent study has given the poor wetting condition, the bubble surface diverges from sphere-like shape due to the movement of gas in the nozzle rim and spread of bubble base at the orifice plate. Davidson and Amick give the empirical relation of bubble volume obtained by experimental results and the majority of models take the assumptions of bubble shape is spherical.

This paper aims to study by using fluent numerical simulation of profile shape effect on bubble creation at a single nozzle underwater at the quasi-static flow rate.

2 Computational Setup

2.1 *Methodology*

The creation of a single air bubble at a nozzle of 0.5 mm radius in a container filled with water and the airflow rate is applied in the nozzle where the bubble formed is kept with fixed volumetric flow rate Q˙ and the bubble evolution studied using a quantity is zero. At the inlet, a parabolic profile of the velocity in the x-direction is axisymmetric geometry as sketched in Fig. 1, where R_0 and D_{eq} are the nozzle/orifice radius and bubble equivalent diameter, respectively. The boundary condition at the centerline is the origin (0, 0) and at the left and right boundary $r = R$ as symmetric boundary conditions, i.e., $u = 0$ and, the gradient normal to the boundary of all specified where $V_0 = Q/(\pi R^2)$ is the mean velocity at the nozzle/orifice. Reynolds number is based upon the flow properties of the gas. We should take the interface of the triple-phase touching line at the orifice plate. Gibbs finds that the triple-phase contact region merges with the point on a solid surface and the contact angle φ is calculated between the gas–liquid transition region where φ_s is the static contact angle of the 3-component system and θ is the angle of the solid wedge angle. In the present case, we take different contact angle with wetting conditions and non-wetting conditions. This means the contact angle $\varphi < \varphi_s$, the bubble base attached to

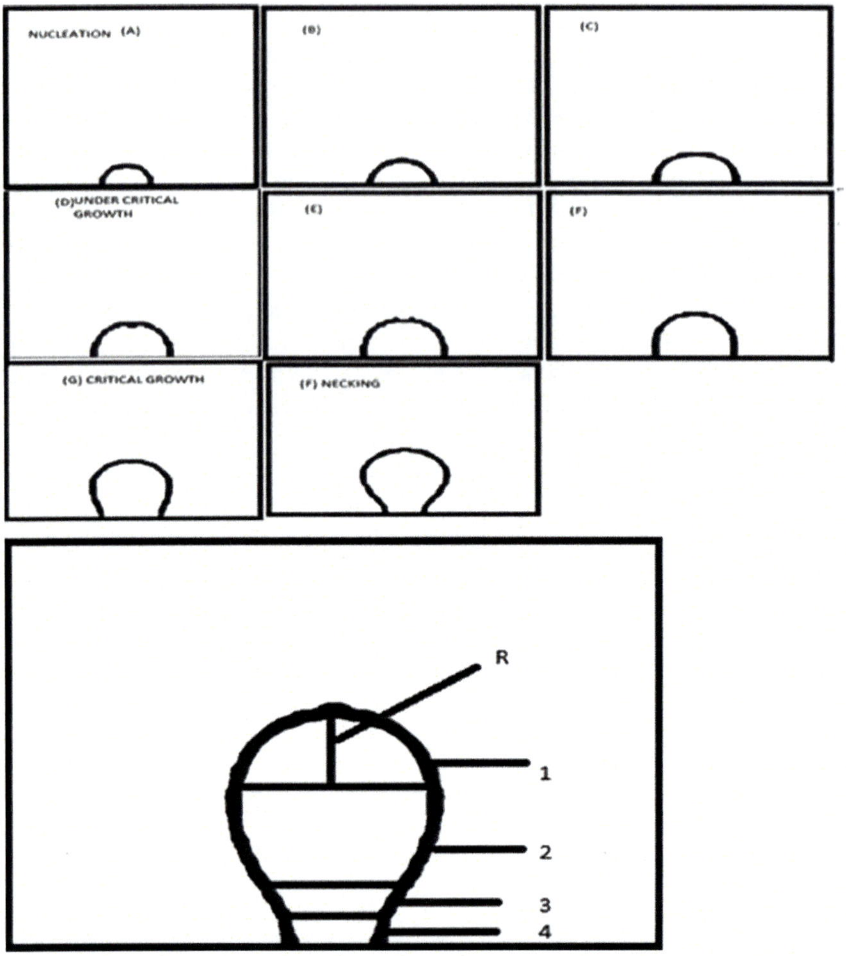

Fig. 1 Different contact angle variation of bubble 0°, 50°, and 110° and different region of bubble

the orifice expands and the moving contact lines manage the constant contact angle $\varphi = \varphi_s$, which does not depend on the speed and the control of the contact line movement. We fixed the contact line for $\varphi > \varphi_s$ and using the fixed contact angle during the contact line movement by fixing the boundary condition $\partial_\varphi/\partial_z = -\cos\varphi_s$. The computational domain describes Fig. 1 as $R = 5\ D_e/2$ and $Y = 10\ D_e$, which is nearly equal to the area $10 \times 20\ \text{mm}^2$ where D_e is the equivalent spherical bubble diameter. The cell size of the mesh is 0.25 mm. This step size is taken by Buwa et al. (2007), Chakraborty et al. (2009), [15]. The time step taken in this study was 1×10^{-6} to satisfy the CFL constraints provided. The time step used to measure the volume fraction was naturally adjusted with defined courant number near the atmospheric surface or outlet, and the relation is as follows:

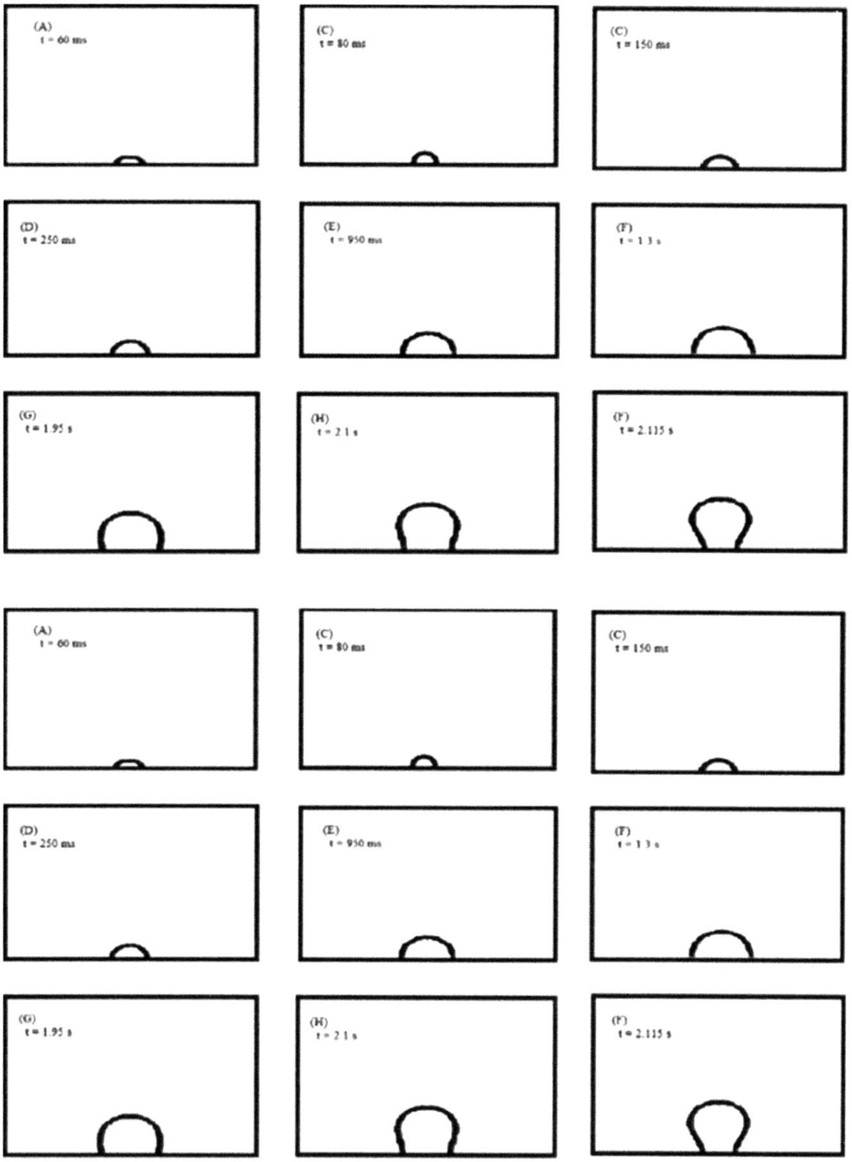

Fig. 1 (continued)

$$N_c = \Delta t / \Delta t^*$$

where Δt is the time step for the calculation of the transport equation and Δt^* is the transport characteristics time of a fluid element of a control volume. The default

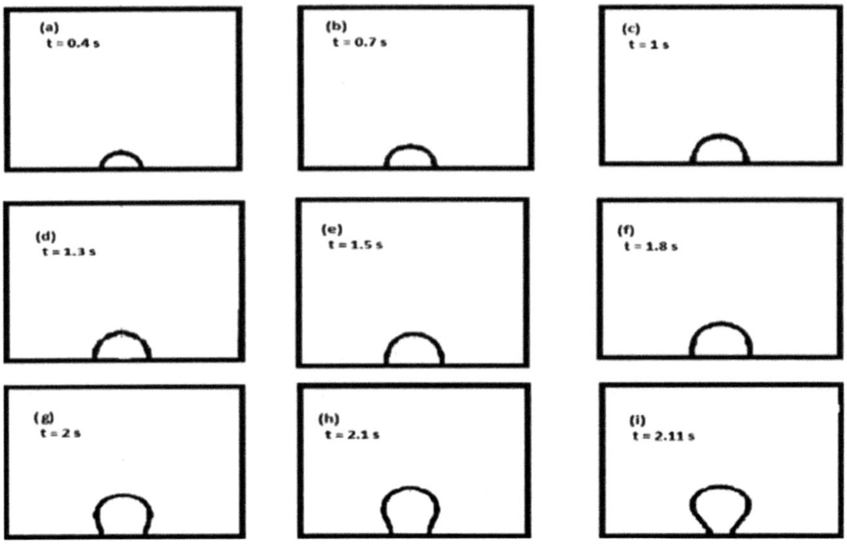

Fig. 1 (continued)

value of the courant number is 0.25. Every bubble shape is broken into geometrical parts: spherical segment, the zone of the sphere, truncated cone, and cylinder for absolute calculation of bubble shape and size. Size is measured for each part of the bubble. The volume (V) of each bubble was calculated from the respective part of the bubble. The dimension of the bubble is calculated more carefully as compared with that of previous studies, which pretend to be a bubble to take the sphere or ellipse-like shape.

2.2 *Phenomenological Description and Measurement Results*

Figure 2 shows the different time step bubble formation at the variation of wetting conditions in the proximity of the orifice.

The key parameter of the bubble is as follow: V, and radius at the tip (R_0), the diameter of bubble periphery at the base (D), contact angle at the triple point line (θ_ω). These specifications are used to identify the forces acting on bubble downwards forces acting on the bubble with upwards buoyant forces. The bubble volume variation clears the transition of one stage to another stage. We check thermodynamic stability from these parameters also and check the energy level of bubble from these parameters. The data we get from numerical simulation regardless of wetting conditions still during each period the bubble evolution shape, volume, and growing movement of the bubble is different. We saw the bubble evolution of each period is defined by using simulation data. The different period predicted the different stages of bubble

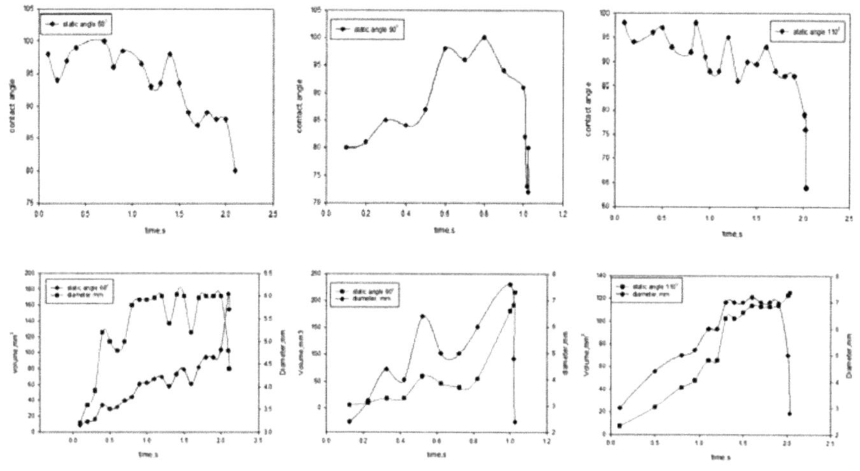

Fig. 2 Volume, diameter change with time, and change of contact angle with time

(1) Nucleation period (2) under critical growth (3) critical growth (4) necking that could be identified easily from the snapshot of the different periods.

2.3 Nucleation Period

The bubble rises from the nozzle in form of a hemisphere-like shape regardless of wetting conditions and it transformed into a hemispherical shape and after that it made a part of a sphere R_0, the radius of orifice and during time-lapse the rising of bubble increase from the orifice and capillary force that act on the bubble is dominating and due to which the nucleation period is larger than the other period. Although the bubble rising from the orifice bubble V increases continuously and V / R_0 increases slightly, the bubble shape becomes stable. Contact angle θ_w decreases since the bubble configuration changes the sphere shape segment which is made of a part sphere. At last, the angle reaches nearly 90° whatever the initial contact angle.

2.4 Stage of Under Critical Growth

In this time, the bubble moves outwards from the orifice and we say that the bubble base moves outwards. The shape of the bubble remains to attach with sphere-like with a center that moves upwards from the orifice. At the poor wetting condition, the shape of a bubble near to sphere-like segment at the edge or upper part of the bubble, which closed to sphere zone at the orifice rim.

In this period, the change of bubble shape occurred. All parameters (V, R_0, D) increase after the nucleation period of the bubble. Bubble shape still sphere-like and geometrical parameter V changed at the start and then smoothly and we calculate that we saw the V/R_0 because this is the stage where buoyancy force just starts to dominate. At this stage, the current contact angle is being varied θ_ω and reaches up to the equilibrium stage. At 90° wetting conditions, the contact angle is constant and it does not differ from the equilibrium value.

2.5 Stage of Critical Growth

In this time, the bubble moves upward and moves outward at the orifice rim. The wettability conditions are good and neutral; the bubble shape is like a sphere-like segment or hemispherical at the edge underneath which a sphere-like section is formed. The truncated cone is formed where the cone apex is downward when the wettability condition is good if the wettability condition is poor then the apex is upward. When the $\theta_0 \leq 90°$, a very negligible cone turned upward, and when the poor wettability condition $\theta_0 > 90°$ the cylinder-like shape which is the part formed between the hemisphere-like tip and trimmed cone moved to apex upwards. The diameter of the cylinder is determined from the hemisphere diameter.

In this stage, the bubble base (D) and orifice diameter (d) and the ratio of these two geometrical properties increase (D/d) which predicts the volume of the bubble increases V. The V increases drastically where $\theta_0 \leq 90°$ and slightly below in $\theta_0 > 90°$ the relation volume increases slightly. At the last of this stage, the V gains the higher dimension size with radius R_0 getting constant values at $\theta_0 \leq 90°$, and if $\theta_0 > 90°$ then the value of V is minimum. When the bubble elongates the bubble periphery attains maximum value and the contact angle increases and gets maximum to this stage. And I found that the V/R_0 gets the maximum value whatever the wetting we take initially. This time instant is the critical stage for bubble growth period where the key parameters get the maximum value.

2.6 Necking Stage

This is the last period of bubble formation; in this stage, the neck formations begin while the bubble takes the proper shape of its curvature and the near to orifice the necking formation begins. At $\theta_0 \leq 90°$, the truncated cone is formed where the apex is upward and above the truncated cone a small cylindrical shape is formed. That's the final shape of the bubble is formed at the necking begins. When $\theta_0 > 90°$ the truncated cone apex is downward, the spheroidization process at the bubble upper part begins and the truncated cone formed at the base turned apex upward.

The necking started then the cylindrical neck formed initially after that the bubble spheroidization occurs after that the bubble neck form which is the diameter at the neck form is more than the orifice diameter whatever the initial wetting condition the bubble size formed at $\theta_0 > 90°$ is greater than the $\theta_0 \leq 90°$.

The diameter of the neck decreases as compared to orifice diameter; at the same instant truncated cone volume increases sharply with height. Neck diameter decreases sharply but it does not attain zero diameter value and its diameter is very less than the orifice diameter. The neck formation is bubble detachment time where the lower truncated one at the base remains attached to the orifice.

2.7 Two Phases of Necking are as Follow

(1) When the neck forms the cylindrical shape is formed before the neck. During this period the base is moving towards the orifice gradually and this time the bubble base does not move far from the nozzle and the truncated cone height remains constant.

(2) At the final stage, the bubble gets detached when the neck narrows towards the orifice. During this time the base move towards the nozzle very sharply and the height of the bubble increases rapidly.

The bubble shape under different wetting conditions at $\theta_0 < 90°$, the bubble volume that keeps a stable growth is reached up to the critical growth period. If we compare the R_0 the value of diameter in necking mode slightly increases than after it is stabilized in strong mode. The bubble volume may be decreased with R_0 still be constant and constant at slight necking mode and after that bubble volume increases at the strong mode. And it applies to different wetting conditions. The surface area S of bubble gets hike at slight mode and become constant at the strong necking mode careless of condition which wetting conditions are taking.

The key parameter bubble varies with slight variation in necking mode, i.e., the volume of the bubble increases. The parameter V/R_0 becomes constant at the wetting conditions $\theta_0 \leq 90°$ and it is decreasing at the $\theta_0 > 90°$ and we see in the simulation during necking mode at the $\theta_0 < 90°$ the volume of bubble becomes maximum. At the necking mode, the curvature of the bubble base (D) and θ_ω value was decreased. During the strong necking mode, D and D/d decrease as we see in the simulation while the θ_ω increases suddenly up to the extreme value.

After the detachment of the bubble, the bubble volume parameter increases due to the spheroidization process developed. After the detachment, the bubble moves upward and aside from the orifice, and less volume of gas leaves on the orifice. Significant change in bubble volume was found different wetting conditions.

3 Experimentally and Theoretical Model's Comparison Between Bubble Evolution

A significant fact between the two models like the volume of bubble measured in the theoretical model is much less than the result got from the experimental. So, bubble volume for high airflow rate can be calculated easily from the given relation by the researcher but for very low airflow rate this relation is not getting satisfactory result. As we study in previous results that the nucleation period and critical growth sage were not taking into a two-stage theoretical model, we consider every step-in numerical simulation. In the two-stage model, focus is on the hemispherical shape only and after that the spherical shape is at good, neutral, $\theta_0 \leq 90°$, and poor $\theta_0 > 90°$ wetting conditions. And experimental is that the transition of air-water of bubble base was detached from the orifice but its diameter is more than the nozzle diameter.

4 Discussion

Bubble evolution variation of contact angle (θ_ω) from equilibrium position (θ_ω) was calculated. We have calculated here contact angle variation when the bubble is evolved in the orifice and the triple contact interaction. In this simulation, we are taking the triple contact phase line where the bubble is evolving from the orifice at a constant flow rate of gas. We are taking the effects of surface phenomena or surface roughness, variation in the composition of chemicals, and movement of particle effect associated with the viscosity of the liquid.

We discuss in this paper the bubble contact angle variation at each time step or different stage.

The nucleation period is that period of bubble where capillary action of force acts with an adhesive force acting on the orifice plate. During this time step, the bubble rises against the action of capillary force and its surface is attached to the orifice plate. At the initial level, the radius of the bubble of the liquid–gas interface contracted due to this pressure in the bubble that increases up to a certain value is attained at the time when bubble becomes hemisphere shape. When the volume of bubble increases then the contact angle θ_ω decreases until the contact angle reaches up to 90° is arrived at the conversation from the nucleation period to transition from the stage of under critical growth. At the nucleation period, the contact angle is static because no movement in the touching line was noticed. Yet, the contact point is forced to shift aside from an orifice because as we injected airflow in the orifice the gas pressure creates a pressure in the bubble. According to Elliot and Riddiford, it gives reasons, why we want to take the static contact angle during the nucleation period, which is referred to as 'receded angle' $\theta_\omega = \theta_R$.

The under critical growth bubble enlarges with a stable sphere-like shape ($\theta_0 \leq 90°$) or hemisphere-like shape ($\theta_0 > 90°$) shape. After that stage the bubble volume increases and pressure decrease that was observed in simulation as well as experimental.

The bubble curvature at the base is increased and the bubble grows at the periphery stably. This means water is displaced by bubble volume and the main factor to displace the water is the surface tension of the periphery of the bubble and different force acting on the bubble periphery and we found that the bubble periphery increases as wetting condition reaches to poor wetting condition.

At $\theta_0 = \theta_\omega = 90°$ the bubble shape is like a dome-shaped cylinder and due to the movement of force is acting on the bubble, the bubble periphery is widening and at the last of this stage, mechanical stability (equilibrium) is achieved. The result obtained from the good ($\theta_0 < 90°$) and poor ($\theta_0 > 90°$) wetting conditions is very difficult to differentiate between them. Since the triple line moves than the spreading of bubble curvature , the contact angle can be characterized like dynamic ones, which is 'receding angle', $\theta_\omega = \theta_R$. At acceptable wetting condition, spreading of bubble surface find difficulties because of the waiting process, which make contact of the triple-phase over orifice plate triple line and it happened due to the receding angle found at the initial of this stage is higher than the static angle $\theta_R = 90° > \theta_0 = 50°$. In non-wetting conditions, the waiting process is highly spreading bubble periphery and due to the receding contact angle found which is smaller than the equilibrium angle $\theta_R \approx 90° < \theta_0$. In any wetting condition, the interface equilibrium is existing when the critical growth stage is near to start. In the critical stage, bubble volume becomes boost still bubble surface remains stable and large. As we see the simulation, the bubble shape is to extend very fast where buoyant force dominates and we know that the buoyant force changes with the volume of the bubble and in this time the bubble height increases without increases the curvature of the bubble and this time step the parameter Volume is maximum because the volume increases sharply. In all the stages, the bubble shape is changed and due to which the bubble parameter is changed and we know the end of the stage, i.e., the necking region where V/R_0 is the maximum value. We see in the simulation picture that the bubble curvature is controlled by the wetting conditions.

As the bubble volume increase, then the pressure will decrease than the triple line contact that becomes weak and the equilibrium will be disturbed, i.e., (metastable stage) and due to which triple line is advanced and static case was appeared during under critical growth and it is referred to as 'advanced angle' rather than the 'receded angle'. As we see in the triple line, displacement was maximum than the 'advanced contact angle' that is arrived at the critical point (volume maximum) before the interface sliding in the necking mode. Elliot and Riddiford proposed that the zero-interface velocity will start to slide. Then in this time step the advanced contact angle appeared.

At the final stage of the bubble, the necking mode of the bubble the buoyant force dominates in this case than the and the force acting on the orifice plate is weakening due to which the bubble moves upwards faster and contact angle variation also occurred. The wetting condition of the bubble predicts the contact angle hysteresis and in the necking mode, the bubble periphery is contracting, and due to which the triple line was making the advanced over the solid surface. Advancing angle $\theta_\omega = \theta_A$, which was larger for equilibrium. In the necking mode, the bubble contact angle is slightly decreased in the initial necking mode and after that the necking of the bubble increases where we said that the disengagement time where the contact angle raises. At the last of critical growth stage, contact angle increases and see that the advancing contact angle relaxes somewhat time with the interface equilibrium as we see in the good and neutral wetting condition without being affected to reduce the contact angle during the initial of necking start. The interface is advancing immediately during the final necking mode as compared to the initial necking mode. This result showed that the 'natural' movement occurred so that the system reverses the move to maintain acted forces an air-water to periphery motion in bump and periphery of bubble step up. At the final mode at better wetting conditions initial and final mode frequently occurred two times twice up to bubble disengagement. The key parameter of bubble V, D, and d is varied in the frequent formation of initial/final necking modes. Finally, spheroidization of bubble uppermost surface and bubble curvature is shrinking towards at the orifice plate and due to which the bubble shape distortion occurs due to which necking and finally neck ruptures. In the above study, the bubble shape is dominated by contact angle hysteresis and force acting on the interface.

Two cases are revealed (1) quasi-static case when triple contact line moved but no movement in the periphery is going on (2) when the dynamic stage is active. Contact angle changes twice that take place during the formation of a bubble. Static hysteresis angle is found in the nucleation period and the stage of critical growth, whereas in the dynamic movement of the bubble under critical growth and necking. At neutral wetting conditions, the dynamic hysteresis occurs at the necking stage, and interface movement was found in under critical growth and necking where the contact angle keeps stable and it was at equilibrium value ($\theta_{\omega} = \theta_0 = 90°$). Static hysteresis is before dynamic. The maximum angle of contact occurs in sliding drop if the periphery is displaced with a jerk.

During the nucleation period, the contact angle must be static and that angle is called the 'receded' angle and similar to the under critical stage the contact angle is almost to be static or we say in a dynamic condition so that this angle is called 'receding', whereas the angle at the critical growth stage and necking are those called 'advanced' or 'advancing angles', respectively. The air–water transition phase is displaced due to change of pressure yet constant flow condition also occurs and we see in the computation result the transition of under critical growth to the critical growth stage, the receding movement is changed so that we ensured advancing the air–water interface. As we take the flow forced flow because we put constant gas flow condition but the transition phase is influenced to get the waiting time of the system back to equilibrium stage that's called the 'natural' displacement. We focus on the critical point where volume becomes extremely maximum. The time at which the balance of force occurs in the direction of detachment. At the final equilibrium interface takes place at air-water inversion displacement take place at different instant times. Finally, we see in the simulation the force is balanced in the detachment time, the contact line is getting disorganized and this perspective of critical growth is the important factor of study and consider each step seriously.

We see in the simulation if the bubble moves upward then the bubble surface tends to detach the bubble nozzle and this time if we equate the forces that we take it is almost balance and we say that the upward and downward force is equal and the adhesive force is minimum at a time to detach but at necking stage, this time force is near to zero hence the negligible volume of the bubble is left in the orifice.

All these computation results predict that the cubic measure of the volume of the bubble depends on the wetting condition of the bubble. Bubble volume is maximum in good wetting conditions as compared to bubble volume in neutral and poor wetting conditions.

5 Conclusion

1 Bubble evolution at a single nozzle (1 mm diameter) underwater with air flowing at a quasi-static flow rate (2 mL/min) takes the static contact angle $500 \leq \theta_0 \leq 1100$).

2 The geometrical parameter (V, R_0, D), two significant stages which is the nucleation stage and growth stage has been studied during bubble growth and when the necking is going to be started and when the initial and final stage of necking is as useful stage of the bubble expansion period.

3 Contact angle hysteresis predicts the bubble dimension variation and two case arises (1) static case, (2) dynamic case. In the first case, no motion is between the transition phase and this time two-stage is defined which is nucleation stage and critical growth stage and the second one is a dynamic case which defined the under critical growth than necking stages. Contact angles are described as 'receded' or 'receding' and 'advanced' or 'advancing' angles are defined as corresponding stages of bubble growth. The transition of air–water surface movement, and force equilibrium in the direction of bubble detachment at the staring time and the final of critical growth correspondingly.

4 This study gives a clear understanding of the bubble expansion of the quasi-static flow rate and gives the different static contact angle.

References

1. L. Davidson, E.H. Amick, Am. Inst. Chem. Eng. J. **2**, 337 (1956)
2. J.F. Davidson, B.O.G. Schüler, Bubble formation at an orifice in a inviscous liquid. Trans. Inst. Chem. Eng. **38**, 335–342 (1960a)
3. A.A. Kulkarni, J.B. Joshi, Bubble formation and bubble rise velocity in gas–liquid systems: a review. Ind. Eng. Chem. Res. **44**, 5873–5931 (2005)
4. R. Kumar, N.R. Kuloor, The formation of bubbles and drops. Adv. Chem. Eng. **8**, 256–368 (1970)
5. J.F. Davidson, B.O.G. Schüler, Bubble formation at an orifice in a viscous liquid. Trans. Inst. Chem. Eng. **38**, 144–154 (1960b)

6. M. Jamialahmadi, M.R. Zehtaban, H. Müller-Steinhagen, A. Sarrafi, J.M. Smith, Study of bubble formation under constant flow conditions. Chem. Eng. Res. Des. **79**(A5), 523–532 (2001)
7. H. Tsuge, Hydrodynamics of bubble formation from submerged orifices. Encyclopedia Fluid Mech. **3**, 191–232 (1986)
8. F.J. Higuera, Injection and coalescence of bubbles in a very viscous liquid. J Fluid Mech. **530**, 369–378 (2005)
9. N.K. Kyriakides, E.G. Kastrinakis, S.G. Nychas, A. Goulas, Bubbling from nozzles submerged in water: transition between bubbling regimes. Cana. J. Chem. Eng. **75**, 684–691 (1997)
10. L. Zhang, M. Shoji, Aperiodic bubble formation from a submerged orifice. Chem. Eng. Sci. **56**, 5371–5381 (2001)
11. A. Tufaile, J.C. Sartorelli, Bubble and spherical air shell formation dynamics. Phys. Rev. E **66**, 056204 (2002)
12. A.B. Ponter, A.I. Surati, Bubble emission from submerged orifices—a critical review. Chem. Eng. Tech. **20**, 85–89 (1997)
13. K. Terasaka, H. Tsuge, Bubble formation under constant-flow conditions. Chem. Eng. Sci. **48**(19), 3417–3422 (1993)
14. V.V. Buwa, D. Gerlach, F. Durst, E. Schlücker, Numerical simulation of bubble formation on submerged orifices: Period-1 and Period-2 bubbling regimes. Chem. Eng. Sci. (2007)
15. D. Gerlach, N. Alleborn, V. Buwa, F. Durst, Numerical simulation of periodic bubble formation at a submerged orifice with constant gas flow rate. Chem. Eng. Sci. (2007)

Design and Development of a Multivariant Exercise Machine (MEM) for the Patients Suffering from Spine-Related Problems

Ruchika Gupta, Sharad Prateek Singh, and Vinod Kumar Yadav

1 Introduction

Chronic low back pain, a prominent reason for ailment in the U.S and other countries, may aggravate problems like sleeping disorder and restlessness [1]. Unrestrained pain, walking inability, and restlessness exhibit poor lifestyle and thus can be treated as an opportunity for doctors to discover effective treatment methods [2]. Non-specific low back pain influences up to 70–80% during their lifetime. In fact, the cause of the pain is certain; however, the pace of recurrence is high and many patients could conceivably prompt inability. Passive treatment may consistently be combined with dynamic treatment to maintain the functional level including work [3]. The common problems encountered in the lumbar region come from muscle strain, tension, or injury in the lower back. The muscles in the lower back feel tightly or constantly cramped. There can be numbness or tingling especially in the feet. The main cause of lumbar pain is usually poor posture, spine injury, and lifestyle [4]. The deskbound behavior while at work aggravates the muscle spines of the lower back. It is advised to opt for frequent breaks between prolonged sitting periods. Further, the activities that are dominant to forestall cardiometabolic and cardiovascular ailments may also be substituted in lifestyle. Roller massage has the potential of becoming an effective means of improving the sufferer's motion characteristic during the idle phase [5]. Most of the studies have revealed that the lower back pains of chronic nature may be effectively treated by following the muscle strengthening exercises. In addition, Motor control exercise (core stabilization) practices, that vary the coordination and control of trunk muscles, are also recommended for the treatment of chronic low back pains. By adopting general exercises, spinal stability diminished due to the optimal activity of lower back muscles [6]. One of the prominent ways of relieving back pain is the Baduanjin exercise, in which eight independent and simple

R. Gupta · S. P. Singh · V. K. Yadav (✉)
G. L. Bajaj Institute of Technology and Management, Greater Noida, Uttar Pradesh, India
e-mail: vinod.yadav@glbitm.ac.in

B. Popuri et al. (eds.), *Recent Trends in Engineering Design*, Lecture Notes in Mechanical Engineering, https://doi.org/10.1007/978-981-16-2900-6_6

movements that advance the mind–body coordination through breath regulation may be used. This further strengthens the abdominal and lower back muscles that ensure better physiological spine curvature [7, 8].

The Chiropractic Manipulative Therapy (CMT) of the thoracic spine, when combined with stretching and strengthening exercises, improves the postural kyphosis in women as spinal manipulative treatment extends the adjoining musculature of the vertebra. This stretch actuates the muscle spindles and Golgi ligament organ reflexes due to which the spinal control will bring about diminished hypertonicity in the muscles, that in-turn adjusts the tone of the muscle [9]. The typical scope of kyphosis is in the range of twenty and forty degrees; however, it may vary with age and gender. The angle of kyphosis increments with age and this expansion is more prominent in females than males [9]. The significantly reclined postures can lead to a decrease in repetitive stress injuries as lying in a fundamentally leaned back pose gives a proportion of alleviation to their lower back pain due to the twisting of the lumbar spine into a progressively neutral posture [10]. Acupressure is a non-invasive, minimal effort, and productive correlative elective clinical way to deal with mitigate pain as it includes use of straining the points situated along the energy apexes of the human body. Acupressure might be considered as a first line non-pharmacologic treatment [11]. Stieglitz et al. [12] reported about the effectiveness of equipment-based pilates convention for reducing the pain, that activates over the deep abdominal and spinal muscles while advancing greater spinal adjustment which improves the pain, minimizes the disability, and improves the evenness of the lumbar multifidus among vertebral level in patients [12]. The risk factors associated with low back pain are several but most of them are due to work-related reasons that include heavy load lifting, forceful motion, twisting, bending, and wrong postural conditions. Low back pain occurs as a consequence of non-occupational risks like age, back injury, etc. [13]. A back rest configuration having less lordotic lumbar posture and lower pressure on the back must be preferred to avoid sitting problems [14]. Exercise interventions focusing on the trunk muscles can be effective for the improvement of low back pain and performing trunk muscle practices at work will decrease the risk of developing pain during standing, particularly when progressing to a sit–stand workstation [15]. Cross-legged sitting postures involve greater kyphotic curves in the lumbar and thoracic spines, thus leading to greater external oblique muscle activation [16]. In the systematic review of pain management techniques, it is found that there is strong evidence to favor multidisciplinary care and stabilization exercises for patients suffering from chronic low back pain [17–20]. Through surveyed literature, it has been observed that the muscle strengthening exercises are primarily the main sources of relieving the lower back pain as induction of steroids and painkillers may severely damage human organs like kidney and liver. To perform various muscle strengthening exercises, as prescribed by doctors, multiple variants of exercise machines are available in markets. However, it is impractical for a common man to afford all these machines due to their high cumulative cost. The specific objectives of the present work were as follows:

(1) To design and fabricate a Multivariant Exercise Machine (MEM) capable of strengthening the weak muscles of patients suffering from spine-related problems and help them to regain their cardiovascular health.
(2) To equip the MEM with adjustable feature and reversing facility (10–60°) as per the needs of a variety of patients.
(3) To provide a paddle exercise facility for strengthening the leg and calf muscles.
(4) To provide heating facilities as per patient's need and doctor's recommendation.
(5) To provide rollers for complete spine massage, in different massaging positions.
(6) To provide rotational facility for stretching of back muscles by twisting the lower back.

2 Materials and Method

2.1 Hospital Survey

The prime objective of the present work is to reduce the back pain and disability in the patients suffering from spine-related and lower back pain problems. A hospital survey is conducted to study the root cause and probable solutions of the spine-related problems. In this regard, a survey of 50 patients at the Indian Spinal Injuries Centre (ISIC), New Delhi (India), was conducted. In addition, the inputs from Dr. Ravinder and Dr. Shiv (Physiotherapist at Roshan hospital, Greater Noida, India) were also recorded. A detailed questionnaire related to the patients suffering from spine-related problems was prepared and the feedback was collected from patients. The common problems suffered by various patients were separated and plotted in Fig. 1, from where it can be observed that the majority of patients suffer from leg pain, lower back pain, and numbness in their feet.

After the detailed analysis of major spine problems, a Multivariant Exercise Machine (MEM) was designed. Through a local market survey, it was observed that the cumulative cost of different machines requires heavy investment. Hence, it has been planned to put all exercising features together in a single machine to

Fig. 1 Hospital survey report

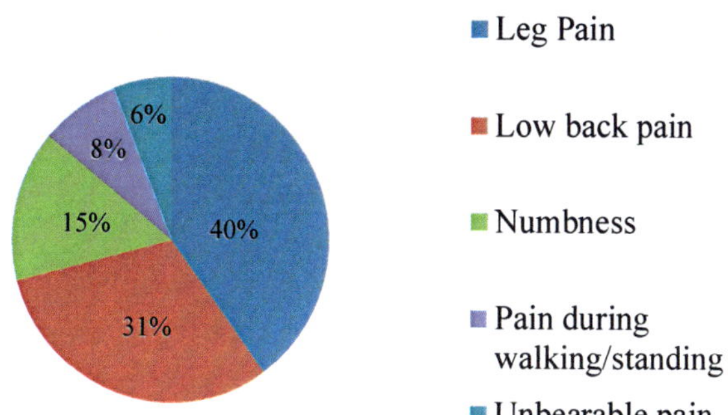

Table 1 Market survey report

Machine part description	Material used	Estimated cost	Source
Inversion Machine	Mild steel and leather for seat	Rs. 27999	Amazon
Heating Pad	Hydroxyethyl cellulose, sodium polyacrylate, or vinyl-coated silica gel.	Rs. 860	Amazon
Cooling pad (Gel)	Sodium polyacrylate	Rs. 429	Amazon
Foot pedal exerciser	Mild steel	Rs. 2027	Flipkart
Back massager with soothing heating facility	Steel and leather	Rs. 16900	Flipkart
Foot stretching machine	Mild steel and spring steel	Rs. 500	Amazon

minimize its cost. The summary of the market survey is presented in Table 1. As can be seen from Table 1, the estimated cost of individual machines when purchased separately will be around INR ₹48286 (INR). Ceragem machine (Make CAREFIT 4500 M: Source www.amazon.com), containing heating and back massager, when purchased separately may alone cost around ₹1, 25,000 (INR).

2.2 *Proposed Design of MEM*

The three-dimensional design of the exercise machine (shown in Fig. 2) was prepared using Solidworks software. MEM can be divided into five basic sections. The chair is the heart of the machine as it houses the roller massage setup and back heating pads. In addition, the chair can also be used as an inversion table. The base of the chair

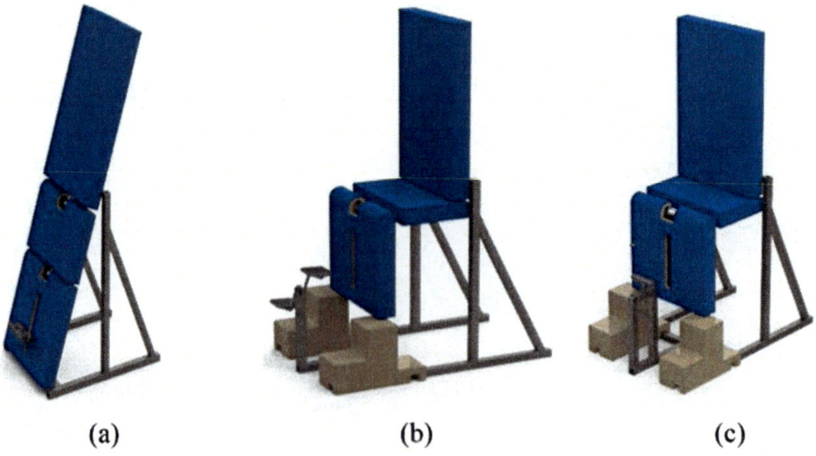

Fig. 2 **a** Inversion table **b** Cycling **c** Weight lifter

carries the mechanism for cycling and the weight lifting. The chair is also equipped with back stretching facility. The main purpose of the back rest is to allow the back and the head to rest. Further, it houses the heating pad that is embedded in the foam layer.

The roller massage mechanism is also fitted. The sub-components of the back rest are shown in Fig. 3. The frame is used to mount all the components housed in the back rest and to support the back and the head of the patient. The wooden load distributor is used to avoid unnecessary load from acting on the roller massage setup and to prevent direct metal contact of heater's wire in case of any damage of the cover. The frame and the wooden load distributors of the back rest are covered by a layer of 20 mm-thick polyurethane flexible foam which has a covering of 2 mm thickness cotton–elastane mix at top and bottom. The foam is also supported by a 3 mm metal wire mesh in between for better rigidity and support. The roller massager will massage the spinal cord of the patient in different modes. It will follow the curvature of the spine. The roller massager is guided on rails and it is driven by a motor connected with timing belts. The lifting of the roller massager is controlled using rack and pinion assembly. The rack and pinion reciprocate up and down to apply the required pressure at the back of the patient. For heating, flexible silicone heaters, embedded in the foam covering the frame of the back rest, are used. The heating pads are controlled by a PID temperature controller for maintaining the desired temperature as per the doctor's recommendation.

The butt rest is simple in construction and is used for supporting the buttocks of the patient. The butt rest, shown in Fig. 4, is connected with the back rest through a pivot which allows for two-axis rotation. The frame is used to support the load of the buttocks via butt rest of the patient. The supporting components of the frame are described in the following paragraph. The wooden load distributor is used to maintain a uniform thickness of the chair (as the thickness of the back rest is higher due to roller massager installed inside the back rest). This also ensures better load

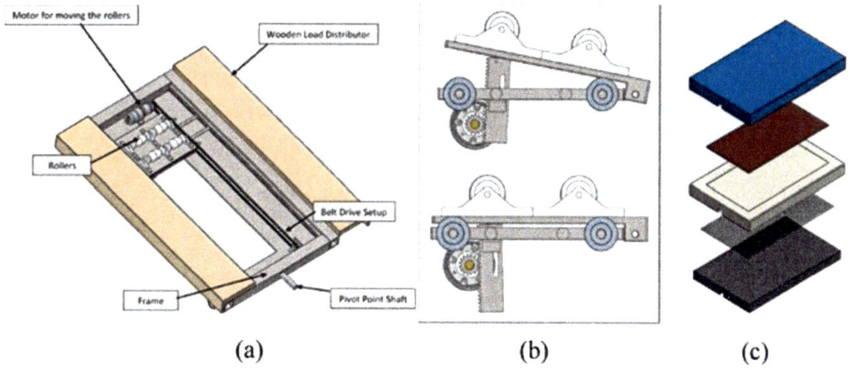

Fig. 3 **a** Back rest components housed **b** Roller massage arrangement **c** Exploded view of foam (Blue: cotton–elastane top cover, red: flexible silicone heater, white: polyurethane foam, mesh: metal mesh for support, black: cotton–elastane bottom cover)

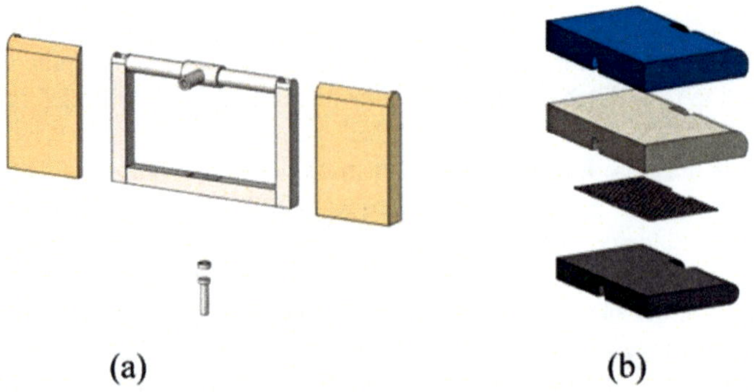

Fig. 4 **a** Components of butt rest, **b** Exploded view of the foam (same as back rest but without heater)

distribution and adds strength to the chair. The frame and the wooden load distributors of the butt rest are covered by a layer of 20 mm-thick polyurethane flexible foam in which a covering of 2 mm-thick cotton–elastane has been fitted at the top and bottom. The foam is also supported by a metal wire mesh in between for better rigidity and support.

The leg rest has a simple construction and is used for supporting the legs and provide a foot mount for the patient. The leg rest, shown in Fig. 5, is connected with the butt rest through a pivot that allows for the rotation of the leg rest. The foot rest is mounted on the frame and has the provision of locking the patient's feet while using the inversion table. The foot rest can be mounted at different positions as per the height of the user. The foot rest can be adjusted to accommodate a height range of 5 feet 2 in.–6 feet 2 in., with an increment of 1 in. per mounting point. The frame

Fig. 5 Components of leg rest

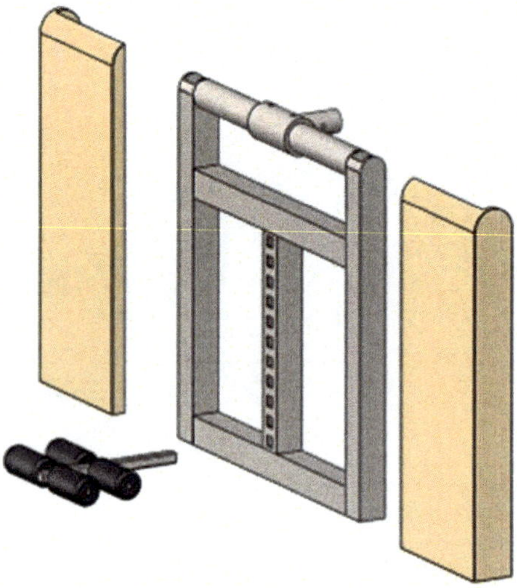

is used to offer a point from which the user can position themselves on the machine and has points to allow the leg rest to be mounted on it. The wooden load distributor is used to maintain a uniform thickness of the chair (as the thickness of the back rest is higher due to roller massager installed inside the back rest). This also ensures better load distribution and adds strength to the chair. The frame and the wooden load distributors of the butt rest are covered by a layer of 20 mm-thick polyurethane flexible foam in which a covering of 2 mm-thick cotton–elastane has been fitted at the top and bottom. The foam is also supported by a metal wire mesh in between for better rigidity and support.

The pivot, mounted between back rest and butt rest, is used to provide the twisting motion required in the lower back for stretching the back, shown in Fig. 6. The heating system for the back is embedded in the foam of the back rest of the chair. This allows user-based temperature adjustment which is controlled by a PID Controller. The roller massage for the back is embedded in the back rest section of the chair and it allows for back massage. The to and fro motion of the rollers is driven by a DC motor that uses timing belts to transfer power. The limiters, used for extremes, are two inductive proximity sensors (Fig. 3b) which sense the metal clip in proximity and stop the motor to limit the motion of the rollers in any given direction (forward or reverse).

The cycling pedal helps in strengthening the calf muscles and ankle. It can work at different resistance levels and the resistance level can be adjusted by turning the knob of the setup which uses friction to vary the resistance on the main shaft as shown in Fig. 7.

The weight lifting arrangement is provided in MEM to strengthen the legs. It has the facility to vary the weight as per individual's capacity and requirement. However, this facility may not be required by the old age patients. Hence, the weight lifting arrangement is made detachable as shown in Fig. 8.

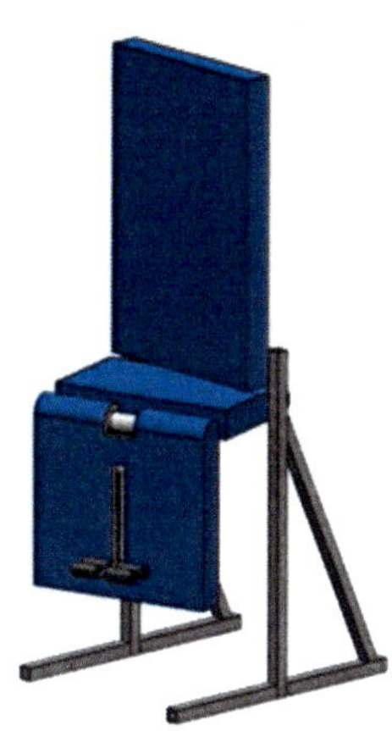

Fig. 6 Rotation of chair's base (butt rest) along the pivot point

Fig. 7 Foot pedal exerciser

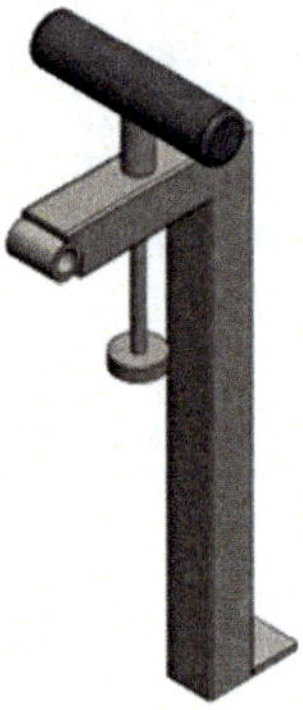

Fig. 8 Weight lifting arrangement

2.3 *Material Specifications*

The materials used for different components of MEM along with their properties are presented in Table 2.

Table 2 List of materials used

Material and its specification	MEM component's name
Mild steel: (AISI 1020) Elastic modulus: $2*10^5$ N/mm^2 Tensile strength: 723.8 N/mm^2 Mass density: 7700 kg/m^3 Yield strength: 620.422 N/mm^2 Shear modulus: 77000 N/mm^2	– Chair frame – Support frame – Pivot point – Shaft of the pivot point – Foot rest
Cotton–elastane blend	– Seat cover of back rest
Polyurethane soft foams	– Cushioning on the chair
Nylon	– Strap

3 Results and Discussions

3.1 Simulation Results

The final design of the exercise machine was simulated to predict its performance and verify that it was in line with the specified design objectives. Simulations were carried using Solidworks software to characterize and optimize the feasibility of the machine. The simulations were conducted for the two conditions in which the chair will work, i.e., as an inversion machine (Table 3) and as a chair for exercising (Table 4).

Figure 9a shows the force applied (purple arrow) on the frame (1200 N). Green arrows show the fixture points of the frame at which various loads will be applied. Figure 9b shows that the frame is discretized and meshed. The software-generated mapped meshing is adopted for static analysis. Figure 9c shows the resultant stress (Von-Mises) as a consequence of applied load. It indicates that the maximum resultant stress (391.851 MPa), for the frame material, is less than the yield strength of the frame material (AISI 1020 with yield strength of 620 MPa). Figure 9d shows the resultant displacement due to the action of the force applied over the frame. It can be concluded that the present design, in its inversion form, is safe as the resultant maximum stress is lower than the yield strength.

Table 3 Results in inversion table mode

Factor	Value
Load	1200 N (shown in Fig. 9a)
Mesh size	4 mm, tolerance: 0.2 mm (shown in Fig. 9b)
Stress	Maximum: 391.85156 MPa Minimum: 0.00000 MPa (shown in Fig. 9c)
Displacement	Maximum: 1.32563 mm Minimum: 0.00000 mm (shown in Fig. 9d)

Table 4 Results for exercising machine

Factor	Value
Load	1200 N (shown in Fig. 10a)
Mesh size	4 mm, tolerance: 0.2 mm (shown in Fig. 10b)
Stress	Maximum: 308.11093 MPa Minimum: 0.00000 MPa (shown in Fig. 10c)
Displacement	Maximum: 1.77919 mm Minimum: 0.00000 mm (shown in Fig. 10d)

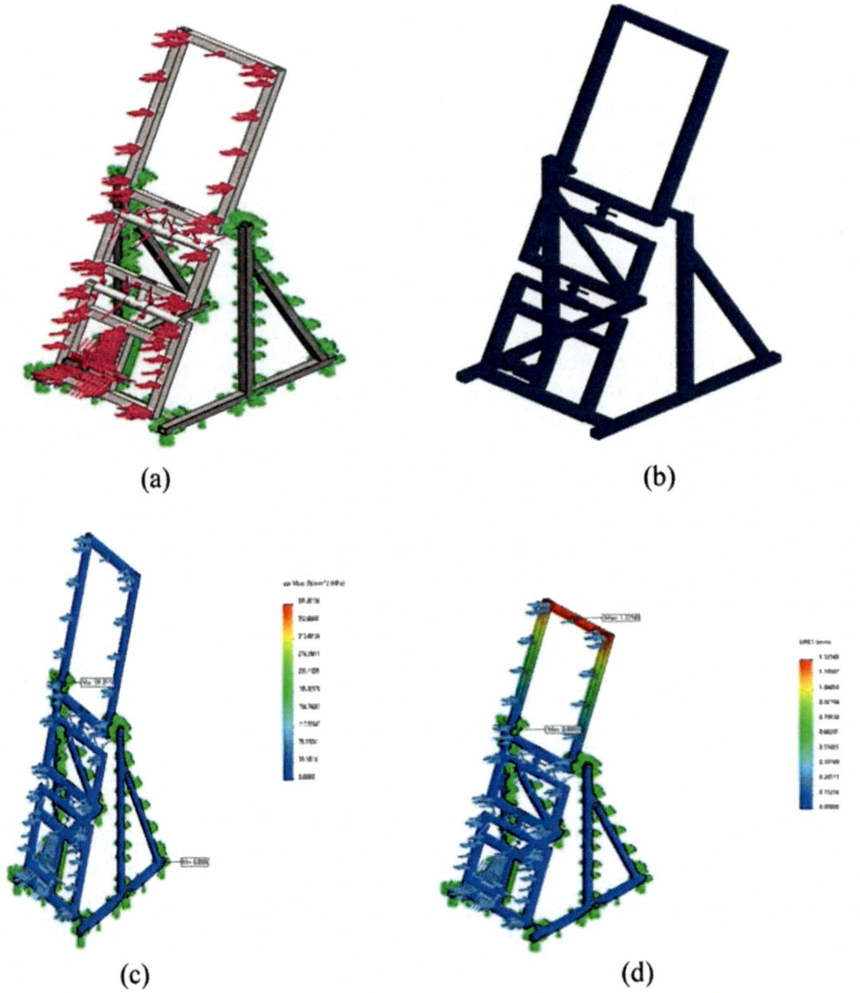

Fig. 9 **a** Loads applied on the frame (purple arrows), **b** Meshed frame of the MEM, **c** Von-Mises stress result of the frame, **d** Resultant displacement from the force applied

Further, the maximum displacement is 1.3 mm, which is safe for the design as it is lower than the maximum allowable displacement (4.5 mm). Figure 10a shows the force applied (purple arrow) on the frame (1200 N) when the machine operates under exercise chair mode. Green arrows show the fixture points of the frame at which various loads are applied. Figure 10b shows that the frame is discretized and meshed, similar to Fig. 9b. Figure 10c shows the resultant stress (Von-Mises) generated due to the effect of applied load. It indicates that the maximum resultant stress (308.11 MPa), for the frame material, is less than the yield strength of the frame material (AISI 1020 with yield strength of 620 MPa).

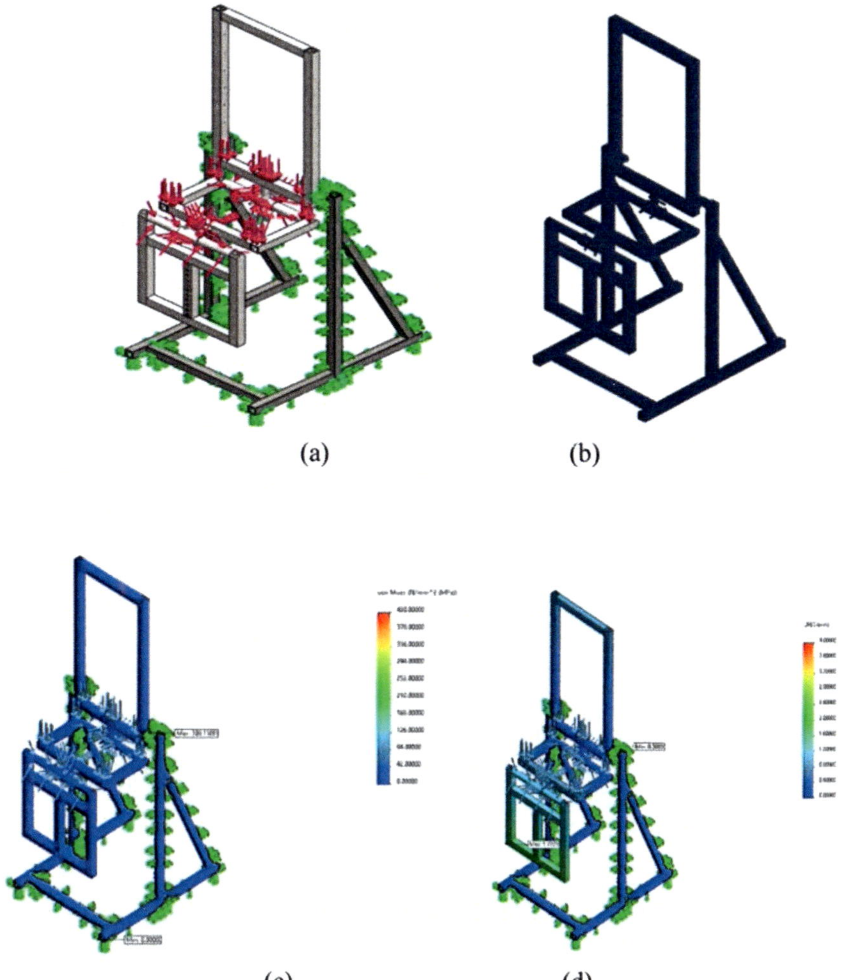
(a) (b) (c) (d)

Fig. 10 **a** Loads applied on the frame (purple arrows), **b** Meshed frame of the MEM, **c** Von-Mises stress result of the frame, **d** Resultant displacement from the force applied

Under this condition, the generated stress is even lower than the inversion mode of the machine. Figure 10d shows the resultant displacement due to the action of the force applied over the frame. It can be observed that the present design, in chair form, is safe as the maximum resultant stress is lower than the yield strength. Furthermore, the maximum displacement is 1.7 mm, which is safe for the design, as it is lower than the maximum allowable displacement (4.5 mm). In this version, it is seen that the maximum displacement is slightly higher than that of the inversion version of the machine. However, it is within safe limits. The results for both versions of MEM (i.e., inversion and chair) show that no substantial deformation under applied load

Table 5 Cost analysis of MEM

Item	Unit price	Quantity	Total cost
AISI 1020	INR 120/kg	83 kg	INR 9960
Foam	INR 275/kg	600 grams	INR 165
SS mesh	INR 14/sqft	6.3 sqft	INR 89
Cotton–elastane fabric	INR 550/kg	900 grams	INR 495
Cycling setup	INR 1600/pc	1 pc	INR 1600
Weight lifting setup + weights	INR 2000/pc	1 pc	INR 2000
Flexible silicone heater	INR 6000/pc	1 pc	INR 6000
PID controller	INR 2000/pc	1 pc	INR 2000
Roller massage setup with belt and pulley	INR 8000/pc	1 pc	INR 8000
Motor	INR 1750/pc	2 pc	INR 3500
SMPS	INR 1250/pc	1 pc	INR 1250
Fabrication and overheads	–	–	INR 5000
Total			**INR 40059**

is observed. Further, the stress generated is 38% less than the yield strength of the material used for frame, i.e., AISI 1020. The resultant stress is about half of that of the ultimate tensile strength of the material. Hence, the factor of safety can be seen close to two. This indicates that the design is safe from permanent deformation as the overall stress generated is lower than the yield strength when tested at maximum capacity (i.e., 122 kg). Although, the frame has been tested without considering the load contribution of sub-attachments that contribute to about 6 kg, but still overall deformation is found to be safe for use due to high factor of safety. The detailed cost analysis is presented in Table 5. It can be seen that the total estimated cost of the MEM is about INR 40059. It can be concluded that the cost is about 80% of the machines available in the market (as mentioned in Table 1 of this paper).

4 Conclusions

In the present work, an innovative design of a Multivariant Exercise Machine (MEM), consisting of multiple assistive features for exercising, was designed and developed. The novel aspects of the design consist of installing adjustable mechanisms, rollers for back muscles massage, cycling, and weight lifting in a single machine. The Multivariant Exercise Machine (MEM) is developed at an affordable cost as compared to multiple individual exercise machines available in the market. This study has provided new insights into the measures which can reduce back pain and prevent spine related issues. The-specific conclusions of the present work can be summarized as follows:

1. The developed Multivariant Exercise Machine (MEM) is capable of strengthening the weak muscles of patients suffering from spinal-related problems.
2. The most effective feature of the MEM is its adjustability (30–100°) that may benefit patients of all age groups.
3. Foot pedal exerciser, as an integral part of MEM, strengthens the weak leg and calf muscles of patients of all age groups.
4. The cost of MEM in some cases is 3.5 times less than the individual machines available in the market, for example, CERAGEM (with back massager arrangement) which costs around INR 1,25,000 (Model CAREFIT 4500 M).

Acknowledgements This work is not supported by any external agency.

Author Contributions All three authors have equal contributions to the finalization of this paper.
Declaration of interest statement
The authors declare no conflict of interest for the present work.

References

1. K.A. Theis, D. Roblin, C.G. Helmick, R. Luo, Prevalence and causes of work disability among working-age U.S. adults, 2011e2013, NHIS. Disability Health J. **11**, 108–115 (2017). https://doi.org/10.1016/j.dhjo.2017.04.010
2. S. Marshansky, P. Mayer, D. Rizzo, M. Baltzan, R. Denis, G.J. Lavigne, Sleep, chronic pain, and opioid risk for apnea. Prog. Neuropsychopharmacol. Biol. Psychiatry **87**, 234–244 (2018). https://doi.org/10.1016/j.pnpbp.2017.07.014
3. S.S. Weiner, M. Nordin, Prevention and management of chronic back pain. Best Pract. Res. Clin. Rheumatol. **24**, 267–279 (2010). https://doi.org/10.1016/j.berh.2009.12.001
4. C. Campbell, S.J. Muncer, The causes of low back pain: a network analysis. Soc. Sci. Med. **2**, 409–419 (2005). https://doi.org/10.1016/j.socscimed.2004.05.013
5. A.R. Kett, F. Sichting, Sedentary behaviour at work increases muscle stiffness of the back: Why roller massage has potential as an active break intervention. Appl. Ergonomics **82**, 102947, (2020). https://doi.org/10.1016/j.apergo.2019.102947
6. M.B. Shamsi, J. Sarrafzadeh, A. Jamshidi, N. Arjmand, F. Ghezelbash, Comparison of spinal stability following motor control and general exercises in nonspecific chronic low back pain patients. Clin. Biomech. **48**, 42–48 (2017). https://doi.org/10.1016/j.clinbiomech.2017.07.006
7. H. Li, D. Ge, S. Liu, W. Zhang, J. Wang, J. Si, and J. Zhai, Baduanjin exercise for low back pain: a systematic review and meta-analysis. Complemen. Therapies Med. **43**, 109–116 (2019). https://doi.org/10.1016/j.ctim.2019.01.021
8. Y. Xie, F. Guo, Y. Lu, Y. Guo, G. Wei, L. Lu, W. Ji, X. Qian, A 12-week Baduanjin Qigong exercise improves symptoms of ankylosing spondylitis: A randomized controlled trial. Compleme. Therapies Clin. Pract. **36**, 113–119 (2019). https://doi.org/10.1016/j.ctcp.2018.12.007
9. K.C. Branco, M. Moodley, Chiropractic manipulative therapy of the thoracic spine in combination with stretch and strengthening exercises, in improving postural kyphosis in women. Health SA Gesondheid, **21**, 303–308 (2016). https://doi.org/10.1016/j.hsag.2016.06.001
10. S. Haynes, K. Williams, Impact of seating posture on user comfort and typing performance for people with chronic low back pain. Int. J. Industr. Ergon. **38**, 35–46 (2008). https://doi.org/10.1016/j.ergon.2007.08.003
11. E. Godley, M.A. Smith, Efficacy of acupressure for chronic low back pain: a systematic review. Complemen. Therapies Clin. Pract. **39**, 101146, (2020). https://doi.org/10.1016/j.ctcp.2020.101146

12. D.D. Stieglitz, D.R. Vinson, M.D.C. Hampton, Equipment-based Pilates reduces work-related chronic low back pain and disability: a pilot study. J. Bodywork Movement Therapies **20**, 74–82 (2016). https://doi.org/10.1016/j.jbmt.2015.06.006
13. F. Gerr, L. Mani, Work-related low back pain. Primary Care: Clinics Office Pract. **27**, 865–875 (2000). https://doi.org/10.1016/S0095-4543(05)70181-0
14. S.M. Carcone, P.J. Keir, Effects of backrest design on biomechanics and comfort during seated work. Appl. Ergon. **38**, 755–764 (2007). https://doi.org/10.1016/j.apergo.2006.11.001
15. V. Johnston, E.M. Gane, W. Brown, B. Vicenzino, G.N. Healy, N. Gilson, M.D. Smith, Feasibility and impact of sit-stand workstations with and without exercise in office workers at risk of low back pain: A pilot comparative effectiveness trial. Appl. Ergon. **76**, 82–89 (2019). https://doi.org/10.1016/j.apergo.2018.12.006
16. S. Ahn, S. Kim, S. Kang, H. Jeon, Y. Kim, Asymmetrical change in the pelvis and the spine during cross-legged sitting postures. J. Mech. Sci. Technol. **27**, 3427–3432 (2013). https://doi.org/10.1007/s12206-013-0865-5
17. D. Pires, C. Caeiro, E. Cruz, Individual patient responder analysis of the effectiveness of a pain neuroscience education programme in chronic low back pain. Manual Therapy **25**, e130 (2016). https://doi.org/10.1016/j.math.2016.05.244
18. S.D. Liddle, J.H. Gracey, G.D. Baxter, Advice for the management of low back pain: a systematic review of randomised controlled trials. Manual Therapy **12**, 310–327 (2007). https://doi.org/10.1016/j.math.2006.12.009
19. I. Caby, J. Vanvelcenaher, A. Letombe, P. Pelayo, Effects of a five-week intensive and multidisciplinary spine specific restoration program in chronic low back pain patients with or without surgery. Ann. Phys. Rehab. Med. **53**, 621–631 (2010). https://doi.org/10.1016/j.rehab.2010.09.006
20. C. Demoulin, S. Grosdent, L. Capron, M. Tomasella, P. Somville, J. Crielaard, M. Vanderthommen, Effectiveness of a semi-intensive multidisciplinary outpatient rehabilitation program in chronic low back pain. Joint Bone Spine **77**, 58–63 (2010). https://doi.org/10.1016/j.jbspin.2009.11.003

Parametric Study and Optimization in Deep Drawing of Headlight Reflector

Jitendra Kumar Singh, Jasleen Kaur, and Rashmi Mishra

1 Introduction

Various parameters are accountable for deep drawing operation, such as punch speed, frictional conditions, blank holding force, properties of blank material, punch radius, die radius, punch–die spacing and thickness of the sheet. Blank holder force, coefficient of friction and punch velocity being the dominant variables were varied in the current study to identify the causes of the cracking and wrinkling.

2 Identification of Problem

Luman Automotive Systems Pvt. Ltd. is a leading producer of a variety of sheet metal products based in Udyog Nagar, New Delhi. The reflector of headlight developed by the company for the Tata 407 vehicle is made with wrinkling and cracking defects via the deep drawing process.

During the drawing operation, the part is found to have the following defects as shown in Fig. 1.

These flawed pieces cannot be assembled in the headlamp of the vehicle since the functions of the headlight are compromised.

J. K. Singh (✉) · R. Mishra
G.L. Bajaj Institute of Technology & Management, Greater Noida, India

J. Kaur
National Institute of Technical Teachers' Training and Research, Chandigarh, Chandigarh, India

B. Popuri et al. (eds.), *Recent Trends in Engineering Design*, Lecture Notes in Mechanical Engineering, https://doi.org/10.1007/978-981-16-2900-6_7

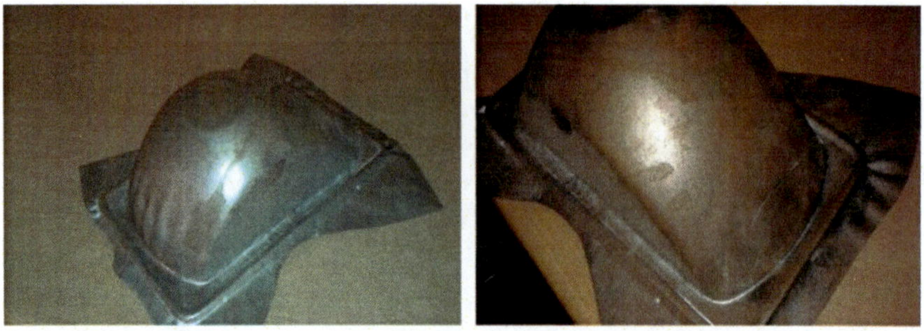

(i) (ii)

Fig. 1 The product having (i) Wrinkling defect and (ii) Cracking defect

3 Experimental Methodology

The various parameters such as punch velocity, coefficient of friction and blank holding force that mainly affect the drawing operation were considered for the study. It was found that the lower value of the blank holding force would result in the development of wrinkles while the higher value would lead to cracking of part. A higher friction force supported to achieve the drawable height of a good amount. Keeping punch velocity value at lower side supported to draw the product by reducing the flow of material, whereas higher value could result in sheet tearing.

The size of the sheet of the workpiece is 330 mm × 310 mm × 0.80 mm. The material of the sheet used was CRCA steel of IS: 513 DD grade.

3.1 Design of Experiments

Three variables, i.e. Blank holding force, Coefficient of friction and Punch speed are taken under consideration. Signal-to-Noise ratio is calculated as per the findings from simulated results. Based upon the objective of the optimization, many ratios are available. The larger the better approach is selected for good quality of the drawing process [2] (Table 1).

Table 1 Selected levels for control variables

Variables	Levels			
	1	2	3	4
A (Blank holding force) (kN)	75	100	125	150
B (Coefficient of friction)	0.01	0.057	0.104	0.15
C (Punch Velocity) (mm/s)	100	150	200	250

3.2 Simulation

The Simufact 11.0.1 application software was used to simulate all 16 experiments as decided by the Taguchi method. According to the L16 orthogonal array, a total of 16 experiments were performed.

4 Analysis of Results

4.1 Calculation of S/N Ratio of Deep Drawing Process

Thickness variation is one of the standards in products formed by sheet metal forming. Thinning is generally responsible for the breakdown of parts made by the deep drawing process; hence, it is favourable to find the strain variation in the direction of thickness simultaneously in the process of deformation. The target is to minimize the change in thickness in products made of deep drawing and also to reduce thinning. Hence, in the present work, thickness distribution is selected as a response derived from experiments. Taguchi's key concept is to regulate indirectly the noise factors by deciding in which way they are influenced by various control factor arrangements. The dual effect of control and noise factor is analysed, and to achieve this, performance criteria suggested is signal-to-noise ratio (S/N) [1]:

$$S/N = 10\log\left(y_m^2/\ s^2\right) \tag{1}$$

$$s^2 = \Sigma(y_i - y_{m)}^2/(n - 1) \tag{2}$$

$$y_m = \Sigma\ y_i/n \tag{3}$$

Here, y denotes the amount of thickness that is under consideration and n denotes the position on the profile of the part where y is measured.

Along the selected profile on the component, the S/N ratios were calculated as shown in Fig. 2.

The evaluated values of S/N ratios, based on Eqs. 1 to 3, for various experiments are listed in Table 2.

Based on the following equations, the calculation of level average and percentage of contribution for each variable is done [1]. Based on all the variation, the overall mean is evaluated is obtained by the following equation:

$$S/N_m = (1/n)\ \Sigma\ (S/N_i) \tag{4}$$

Here, n indicates the total number of tests performed.

An overall sum of squares is provided by

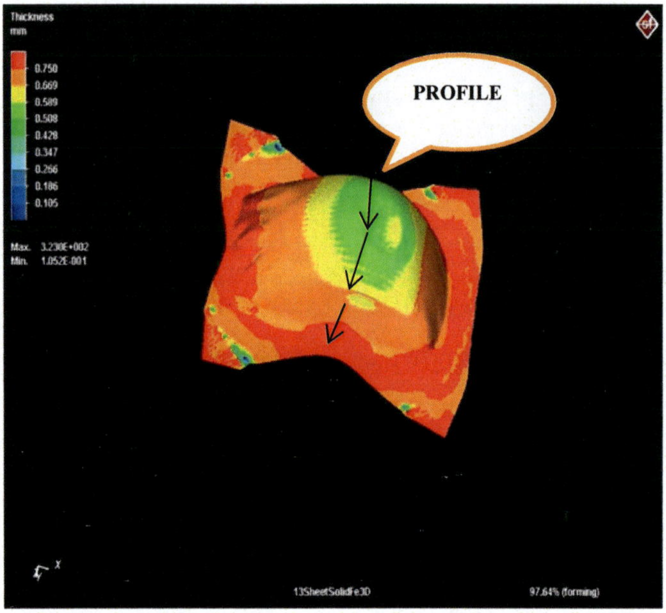

Fig. 2 Values of thickness taken on selected component profile

Table 2 Values of S/N ratios

Test no.	S/N ratio	Test no.	S/N ratio
1	16.342057	9	15.367611
2	16.48196	10	16.859672
3	15.712359	11	15.918606
4	15.909647	12	15.133874
5	14.892450	13	15.757872
6	16.225865	14	16.030342
7	15.943262	15	17.482674
8	16.251670	16	15.692865

$$S_T = \Sigma\,(S/N_i)^2 \tag{5}$$

which is to be divided further into two components; the sum of squares for total mean and for the variation around total mean:

$$S_T = S_m + S_v \tag{6}$$

The sum of squares for total mean is

$$S_m = n.(S/N_m)^2 \tag{7}$$

Here, n denotes the total no. of experiments. The sum of squares for variation around the total mean is

$$S_v = (1/n)\,\Sigma(S/N_i - S/N_m)^2 \tag{8}$$

The S_v is now divided into the sums of the squares of the variation produced by each variable affecting the total mean. For variable A, the sum of squares for variation about total mean is

$$S_{jA} = n_{A1}(S/N_{mA1} - S/N_m)^2 + n_{A2}(S/N_{mA2} - S/N_m)^2 + n_{A3}(S/N_{mA3} - S/N_m)^2 \tag{9}$$

In this, n_{Ai} represents values n of experiments performed at level i of variable A; S/N_{mAi} denotes level average S/N of variable A at level i. Likewise, the sum of squares because of variation about total mean is evaluated for the rest two variables. Then, the participation of each variable is evaluated as follows:

$$C_j = S_{jA}/S_v \tag{10}$$

The average value of S/N ratio in all tests is provided as 18.32. The level mean response analysis due to S/N ratios is provided in Fig. 3 (Table 3).

The level which provides the largest S/N ratio was selected, for optimized values of the selected variables [1]. Hence, it can be determined that the optimum level for Blank holder force, Coefficient of friction and Punch Speed is 150 kN, 0.057 and 150 mm/s, respectively. The values of S_v, S_j and C_j are evaluated on the basis of Eqs. 8, 9 and 10. The calculated value of S_v comes out to be 2.540642. The following table shows the calculated values of the contribution of each variable.

This is concluded that the Coefficient of friction is having the highest contribution, i.e. 72.65%, and the blank holding force and punch velocity are accounted for the lesser contribution.

The Coefficient of friction has a major influence on the quality of the drawn part having a percentage of 73%. That is followed by the Blank holding force having a percentage of 21%. The Punch Speed was having the smallest contribution having a percentage of 6%.

4.2 *Results Obtained by Simulation of Deep Drawing Process by Applying Optimum Variables*

The simulation was carried out by applying the optimized values of variables as provided in Table 4. The variation of thickness after the completion of the simulation process on the basis of optimal variables and current variables is displayed in Fig. 4.

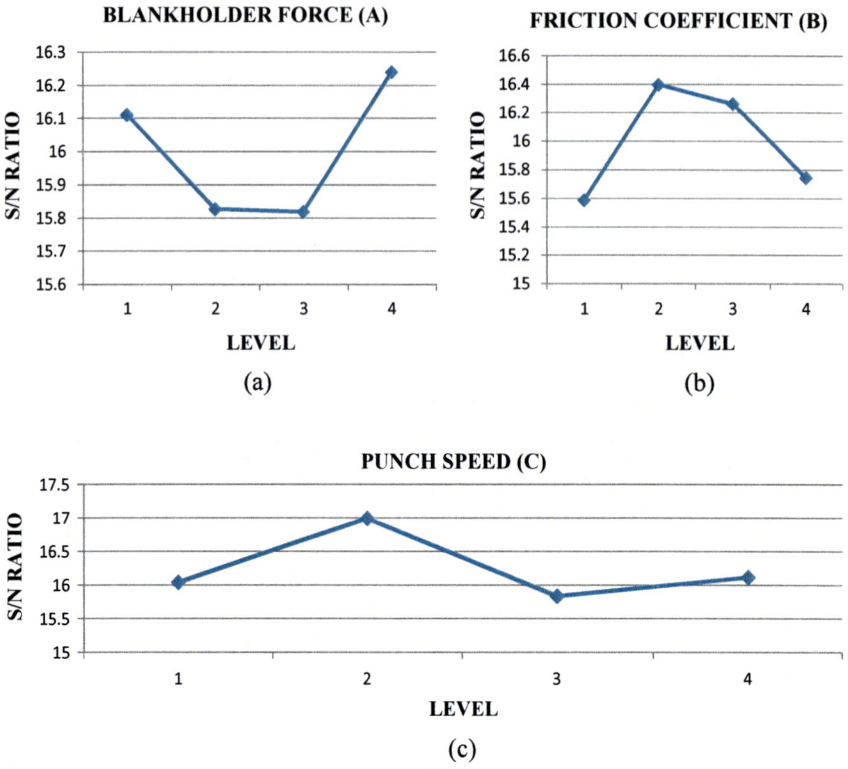

Fig. 3 Graph of average values of levels for selected variables **a** Blank holding force; **b** Coefficient of friction and **c** Punch speed

Table 3 Comparision of actual values with optimum values

Control factor	Actual level	Optimum level
A (BHF)	100 kN	150 kN
B (Friction Coefficient)	0.01	0.057
C (Punch Speed)	200 mm/s	150 mm/s

It was estimated that the wrinkled area in the curved portion that is inclined had decreased relative to the actual case. It was also found that the thickness variance with optimum variables was 0.65 mm to 0.75 mm in the wrinkle-prone region; however, it remained 0.66 mm to 0.86 mm by using the current variables. It suggests an increase in the variation of thickness. Hence, it was estimated that by using optimized parameters, the wrinkles could be reduced. The thinning in the surrounding area nearby nose radius was increased however the lowest value of thickness was 0.512 mm, which was still greater than the previous existing value of 0.293 mm.

Table 4 Contributions of variables

Parameter	Level	S/N_{mn}	S_j	% Contribution
A (BHF)	75 kN	16.111506	0.529532	20.842448
	100 kN	15.828312		
	125 kN	15.819941		
	150 kN	16.240938		
B (coefficient of friction)	0.01	15.589997	1.845946	72.656674
	0.057	16.399459		
	0.104	16.264225		
	0.15	15.747014		
C (punch speed)	100 mm/s	16.044848	0.1651639	6.500872
	150 mm/s	15.997739		
	200 mm/s	15.840495		
	250 mm/s	16.117613		

5 Outcomes

The following outcomes are obtained after analysing the data obtained and explained as follows:

(i) It has been estimated that blank holding force, coefficient of friction and punch speed were identified as the most dominant parameters which are affecting the drawing operation.

(ii) The contribution of coefficient of friction was found to be 73% as far as the excellence of the drawn component is concerned. The optimized value was increased to 0.057 as compared to the actual value of 0.01. Hence, it was concluded that the flow of material can be restricted by increasing the value of friction coefficient, as a large amount of metal flow into the die cavity leads to thickening of sheet and so wrinkles are induced.

(iii) The contribution of blank holder force was found to be 21%. The suggested value is 150 kN as compared to the original value of 100 kN. This implies that to have proper metal flow while drawing, the blank holder value must be increased.

(iv) The contribution of punch speed was found to be 6%. The suggested value is 150 mm/s as compared to the original value of 200 mm/s to obtain good quality metal deformation.

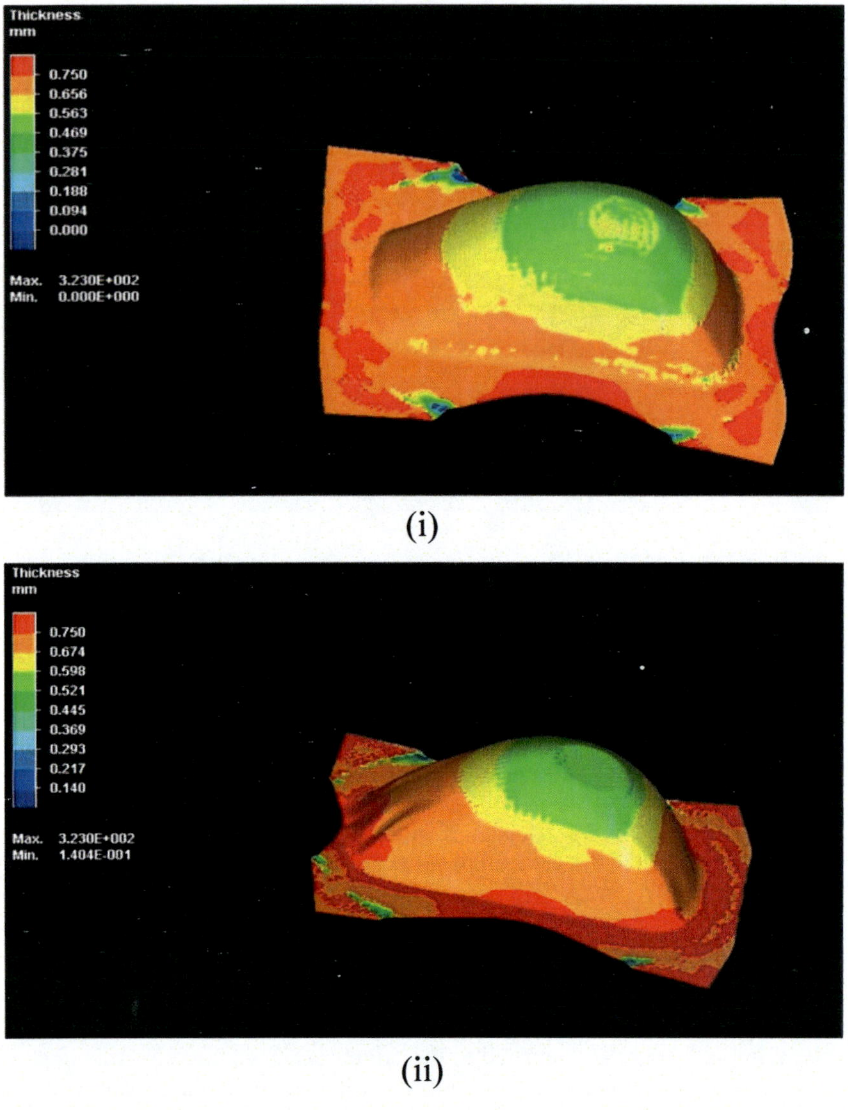

Fig. 4 Variation of thickness during (i) Optimal variables and (ii) Current variables

References

1. S. Raju, G. Ganeshan, R. Karthikeyan, *Influences of variables in deep drawing of AA 6061 Sheet* (Trans. Nonferrous Metal Soc., China, 2010), pp. 1856–1862
2. T.P. Bagchi, *Taguchi methods explained* (Prentice Hall of India, New Delhi, 1993)

Numerical Simulation of a Race Car Wing Operating in the Wake of Leading Vehicle with Varying Diffuser Angles

C. Animesh Bharadwaj, V. V. S. Ganesh, and B. T. Kannan

1 Introduction

One of the most significant advances which has picked up consideration of automotive producers around the globe is the aerodynamic features of a Formula One race car. Indeed, even the smallest of changes in the aerodynamic features of race cars can prompt a huge change in execution. Also, a race car must convey high performance in different factors so as to decide how fast a vehicle can race in the circuit. The basic desired function of the aerodynamic components of a race car is to minimize the drag and maximize the downforce. The current study shows the aerodynamic efficiency of a wing operating in a wake, where an upstream bluff body incorporating a diffuser has been altered.

The diffuser creates nearly 50% of the car's downforce. The air gushing underneath the vehicle is vented out from the diffuser of the vehicle located at the rear end. Although wings and diffusers work comparably similar, they are based to serve various purposes. A diffuser serves to eject air out from the underside of the car. Due to the pressure differences generated between the top part of the car and below, a downforce is formed and hence is the car pushed towards the ground. The suction caused by the ground on the vehicle is a result of Bernoulli's equation, which states that the increase in velocity of a fluid is associated to the decrease in pressure. Accordingly, the velocity of the air stream below the race car would be greater than the velocity at the exit if the pressure below the race car must be lower than the pressure at the exit. The main aim of the diffuser is to widen the flow from beneath the vehicle to the rear and reduce the speed of the flow from the diffuser's inlet to the exit. The front wings provide around one third of the overall downforce provided by the entire vehicle. Unlike the rear wing, the front wing does not only produce

C. Animesh Bharadwaj · V. V. S. Ganesh · B. T. Kannan (✉)
Department of Aerospace Engineering, SRM Institute of Science and Technology, Kattankulathur, Chennai 603203, Tamil Nadu, India

B. Popuri et al. (eds.), *Recent Trends in Engineering Design*, Lecture Notes in Mechanical Engineering, https://doi.org/10.1007/978-981-16-2900-6_8

downforce. It is also responsible for directing the airflow back to the rest of the car, as it is the aerodynamic system that precedes the entire car.

2 Literature Survey

Simon Durrer et al. have worked with the race car wings. In this thesis, the wing was studied in the wake region of the bluff body; prior to this, the wing was analysed in the freestream and ground effect. Later, the study involved different ground clearance levels, and the lift and drag coefficients were calculated for the front wing in the wake region [1].

Aerodynamic factors that may influence the development of overtaking opportunities in open-wheeled racing categories such as Formula 1 in the paper were studied by Soso and Wilson [2]. They experimentally studied the effect of various diffusers with varying angles and the ground clearance on the wing in the wake region. The smallest diffuser slope increased the performance of the downstream wing which resulted in generation of vortex wake generation of the diffuser which also resulted in wake characteristics of the wing.

In Mokhtar and Durrer [3] the high lift airfoil S1223 was studied to understand the aerodynamic forces in ground effect. At multiple ground clearance levels, the wing was experimentally tested and it was understood that the downforce is increased when the ground clearance is reduced.

In Kachare [4] Formula one cars use multi-element front wing to increase the efficiency, and the characteristics of this wing have been studied using three-dimensional computational analysis. The simulation was conducted to see the behaviour of the downforce generated with respect to ground clearance. It was again understood that the ground clearance affects the downforce generated. They have used the bluff body with open wheel, and the downforce on the multi-element analysis is studied by placing it in the wake of the bluff body.

In Andrea [5] by the force and pressure were studied on a basic bluff body with 17° diffuser and side plates. The diffuser characteristics were studied in ground effect with the implementation of the side plates.

To simulate a vehicle following another, a simple single-element front wing is kept in the wake region of the car which is being followed. The bluff body in the front has been incorporated with a diffuser and rear wing. The elevation and angle of attack of the wing were varied and the results were analysed. The wing of the body placed in front experienced a decline of downforce, the loss was due to the variation in ground clearance [6].

In Andrea [7] Generic bluff body with a diffuser ramp equipped with side plates in ground effect was studied to understand the force and pressure characteristics. Multiple diffuser angles were tested, 5, 10, 15, 17 and 20 degrees, in a low-speed wind tunnel with ground effect. The lower angle diffuser was found to be more efficient.

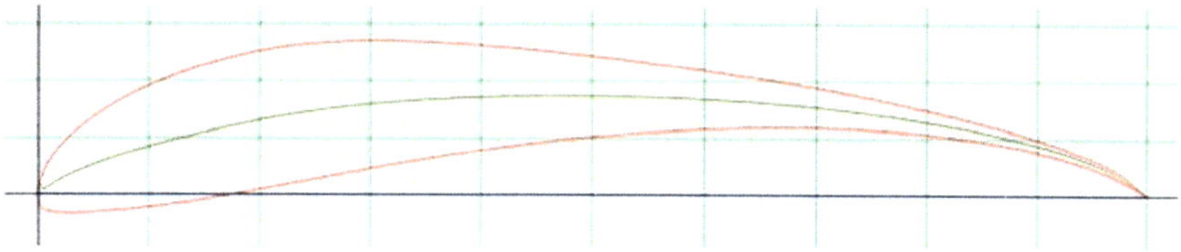

Fig. 1 S1223 airfoil plot

In Moore et al. [8] both experimental and computational methods are incorporated to understand the aerodynamics of wings in ground effect. The Laser Doppler Anemometry and Particle Image Velocimetry techniques are used to read the wake of the wing at the centre region and the behaviour of the tip vortices of the wing. The gurney flaps were added and the change in flow was observed.

3 Project Methodology

By using ANSYS Fluent, the study attempts to obtain the results by doing a computational analysis. After the selection of airfoil, CATIA is used to make the airfoil. Design modeller in ANSYS has been used to make the domain around the model and is validated. The parameters for boundary conditions are made after the selection of the solver. The post processing part involves visualizing the contours, velocity distribution and streamlines. The data acquired is tabled for a better comparative understanding of the results.

4 Selection of Airfoil

There are various types of airfoils. Choosing a suitable airfoil is very important; the airfoils are chosen based on the objective and what we want to achieve. A high lift airfoil has been used to produce high downforce and increase the performance of the race car. The front wing and rear wing (spoiler) use the same airfoil. The airfoil chosen is S1223 as the downforce produced matches the requirement [1] (Fig. 1).

5 Design Specifications of Bluff Body with Chosen Diffuser Angles

The diffuser angles were chosen from the literature survey, these three angles 5°, 10° and 16.7° were selected for the study [7].

Fig. 2 16.7° diffuser angle

Fig. 3 10° diffuser angle

Fig. 4 5° diffuser angle

6 Front Wing Placed in the Wake Region

The chord length of the front wing is 220 mm which is placed behind the bluff body which is incorporated with a spoiler as shown in the figure below.

The domain for the three diffuser angles is created on ANSYS Workbench design modeller (Figs. 2, 3 and 4).

7 Validation

The airfoil is validated at 1,000,000 Reynolds number (Tables 1 and 2).

8 Mesh Convergence Study

See Table 3 and Fig. 5.

Table 1 Validation of C_l versus α

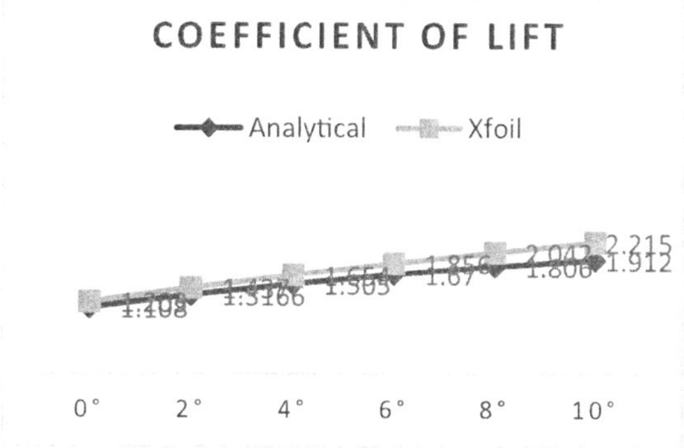

Table 2 Validation of C_d versus α

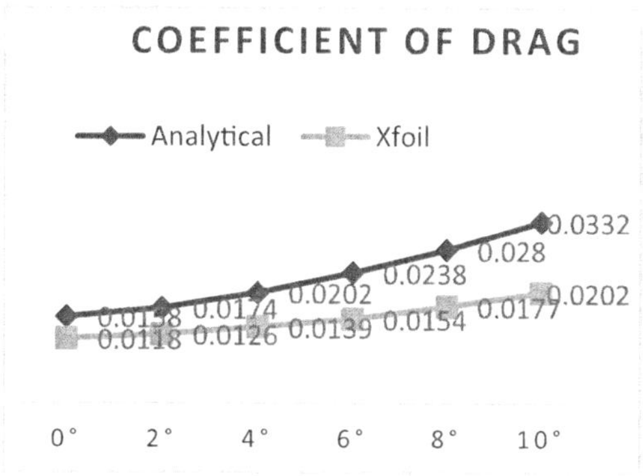

Table 3 Mesh convergence

Mesh number	Downforce (N) (not to be scaled)
Mesh 1	6.775
Mesh 2	6.429
Mesh 3	6.362
Mesh 4	6.263
Mesh 5	6.262

9 Boundary Conditions

The Airfoil is enclosed which acts as the medium. Reynolds Number = 500,000. The velocity can be calculated using the relation = / where $\mu = 1.4207 \times 10^{-5}$ Ns/m^2 and = 1 kg/m^3. The simulations were performed at two velocities, 45 m/s and 60 m/s.

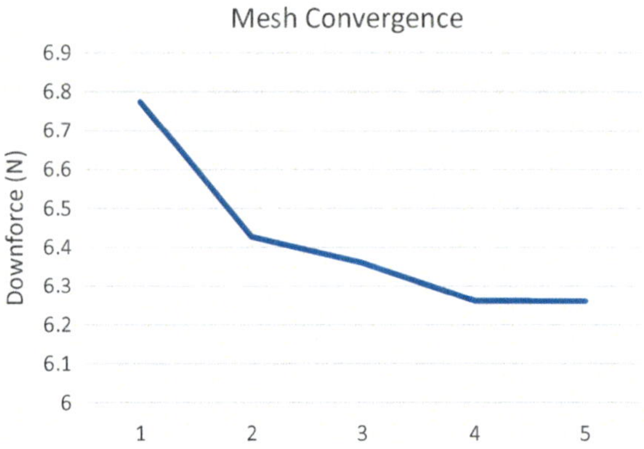

Fig. 5 Mesh 4 is chosen for analysis as it is found to be the ideal one

1. The enclosure which acts as the fluid is meshed and then the faces are named. The leading-edge side wall is named as "Inlet" to which we give *Velocity Inlet* boundary condition where the inlet velocity flows in the X-direction.
2. The trailing edge side wall is named as "Outlet" to which *Pressure Outlet* boundary condition is given.
3. The top and bottom edge walls of the domain are considered as *walls*.

10 Numerical Model Conditions

The model is analysed using ANSYS FLUENT. The viscosity model used in the simulation is **K-epsilon (2 eqn.) realizable** as it shows high accuracy for boundary layers under high pressure gradients and flow separation [9]. The fluid characteristics are changed from default to ideal gas characteristics, the density is considered as 1 kg/m^3 and the viscosity is considered as 1.4207×10^{-5} Ns/m^2. The boundary conditions remain unchanged for every case as the objective is to understand the difference in performance for different diffuser angles at two chosen velocities. The downforce monitor is added to the front wing placed behind the bluff body to understand the variation in each case.

11 Results

The increase in downforce in each case is compared with the bluff body without a diffuser (Tables 4 and 5).

The 5° diffuser angle was found to be the most efficient in both the cases. While 16.7° was marginally more efficient than 10° in both cases, the substantial improvement in efficiency of the lower angle diffuser can still be seen.

Table 4 Downforce of different diffuser angles at 45 m/s and 60 m/s

Diffuser angles	Downforce (N)	
	45 m/s	60 m/s
5°	6.142	11.245
10°	5.789	10.501
16.7°	5.88	10.539
0°	5.273	9.464

Table 5 Comparison of three diffuser angles at 45 m/s and 60 m/s

Diffuser angles	Increase in Downforce	
	45 m/s (%)	60 m/s (%)
5°	16.10	18.82
10°	9.78	10.95
16.7°	11.50	11.35

12 Result Visualization

Inlet Velocity −45m/s.

See Figs. 6, 7, 8 and 9.

Inlet Velocity −60m/s

See Figs. 10, 11, 12 and 13.

Turbulence contours at 45m/s

See Figs. 14, 15, 16 and 17.

Fig. 6 Velocity distribution for diffuser angle −5

Fig. 7 Velocity distribution for diffuser angle $-10°$

Fig. 8 Velocity distribution for diffuser angle $-16.7°$

Fig. 9 Velocity distribution for diffuser angle $-0°$

Fig. 10 Velocity distribution for diffuser angle −5°

Fig. 11 Velocity distribution for diffuser angle −10°

Fig. 12 Velocity distribution for diffuser angle −16.7°

Fig. 13 Velocity distribution for diffuser angle −0°

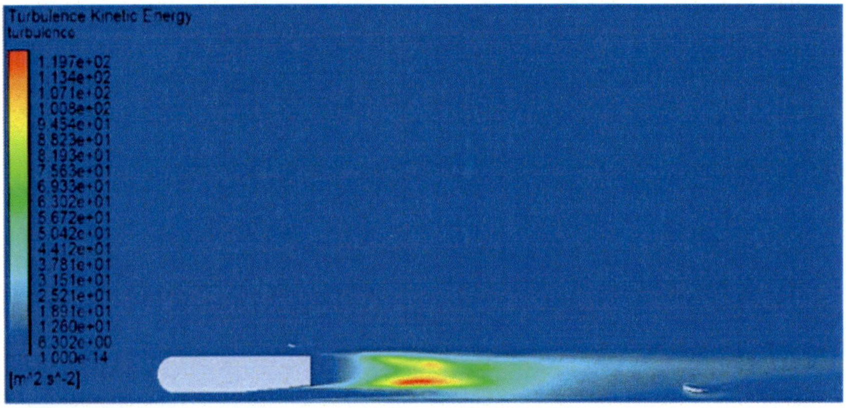

Fig. 14 Turbulence contours for diffuser angle −5°

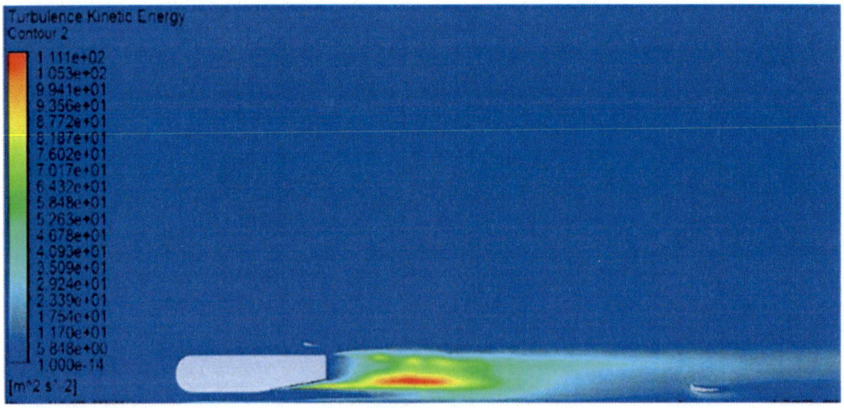

Fig. 15 Turbulence contours for diffuser angle −10°

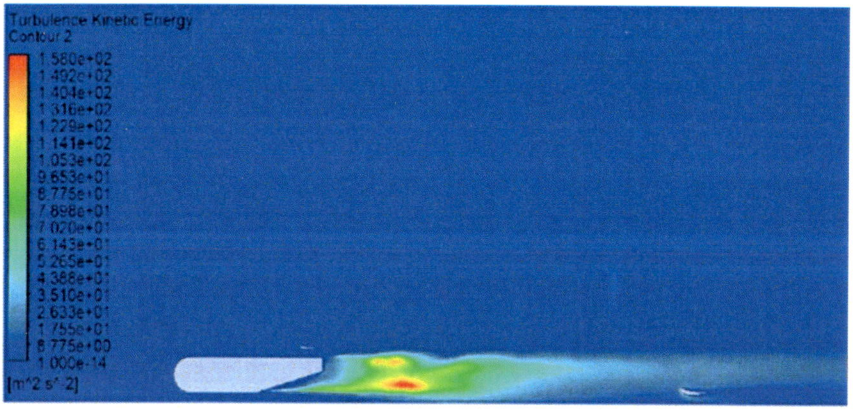

Fig. 16 Turbulence contours for diffuser angle −16.7°

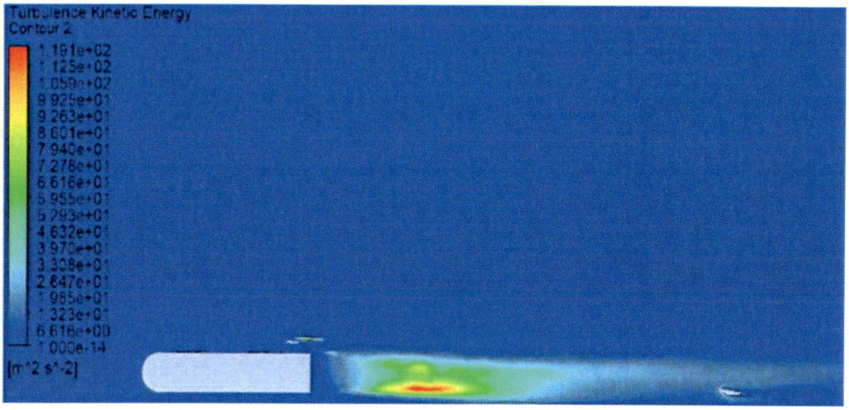

Fig. 17 Turbulence contours for diffuser angle −0°

Turbulence contours at 60m/s

See Figs. 18, 19, 20 and 21.

Velocity streamlines at 45m/s

See Figs. 22, 23, 24 and 25.

Velocity streamlines at 60m/s

See Figs. 26, 27, 28 and 29.

Fig. 18 Turbulence contours for diffuser angle −5°

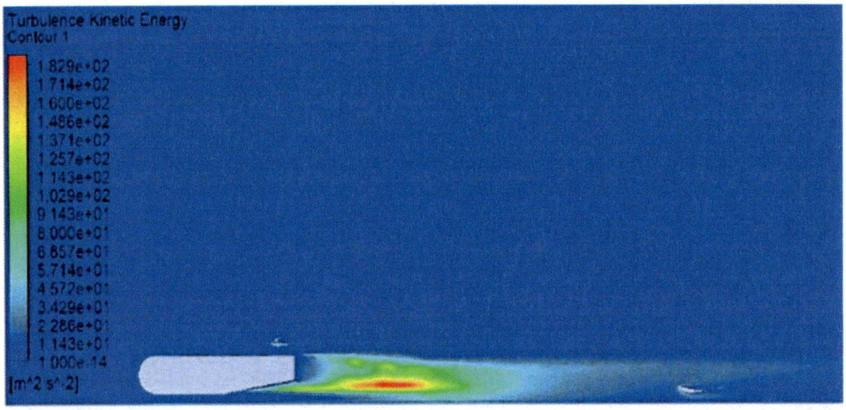

Fig. 19 Turbulence contours for diffuser angle −10°

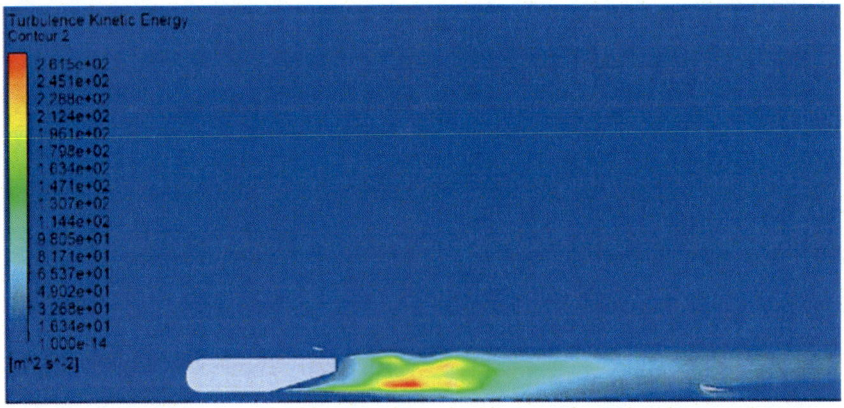

Fig. 20 Turbulence contours for diffuser angle −16.7°

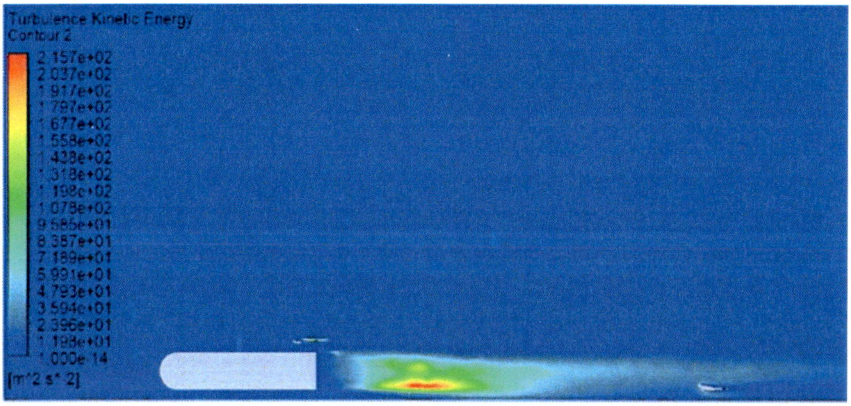

Fig. 21 Turbulence contours for diffuser angle −0°

Fig. 22 Velocity streamlines for diffuser angle −5°

Fig. 23 Velocity streamlines diffuser angle −10°

Fig. 24 Velocity streamlines for diffuser angle −16.7°

Fig. 25 Velocity streamlines for diffuser angle −0°

Fig. 26 Velocity streamlines for diffuser angle −5°

Fig. 27 Velocity streamlines for diffuser angle −10°

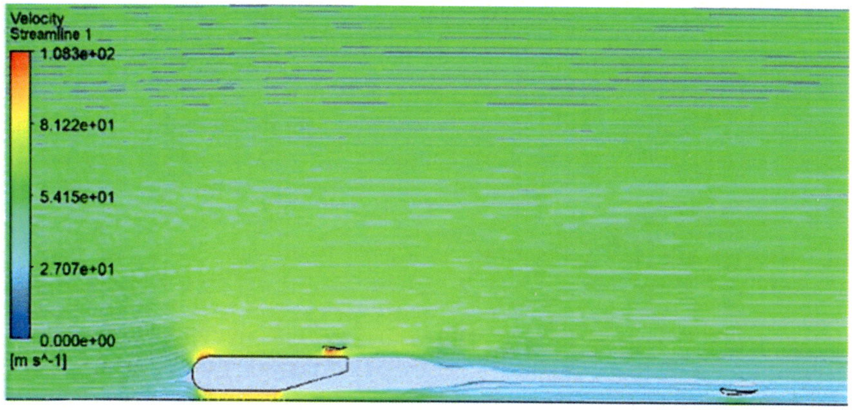

Fig. 28 Velocity streamlines for diffuser angle −16.7°

Fig. 29 Velocity streamlines for diffuser angle −0°

13 Conclusion

Several conclusions can be made from the results we have obtained. The analysis of a single-element wing placed in the wake of the diffuser gives us a good understanding of the aerodynamics forces influenced by various diffuser angles. The size of the wake is not only controlled by the size of the bluff body but it is also controlled by the velocity [1]. Hence, the following conclusions have been outlined from the acquired results.

1. Bluff bodies with diffusers and without diffusers have had comparable downforce levels on the vehicle. The ones with the diffusers have delivered more downforce and reduced the flow velocity under the vehicle.
2. Varying the diffuser angles has influenced the downforce levels on the downstream wing. An increase in downforce on wing place in the downstream was observed as the diffuser angle was decreased.
3. The 5° diffuser angle was found to be the most efficient in both cases. While 16.7° was marginally more efficient than 10° in both cases, the substantial improvement in the efficiency of the lower angle diffuser can still be seen.
4. The flow separation which happens in the diffuser region in the higher angles can be observed in the turbulence contours.

14 Future Work

The present project can be explored with various other possibilities. Since the study of the wake region of a race car is limited, the wing could now be analysed with various aerodynamic components such as gurney flaps. Furthermore, the study can be performed using various offset distances and different angles of attack for the front wing. It would be intriguing to carry out a 3D simulation considering a multi-element wing and different airfoils. Additionally, taking into account that the wing is influenced by the wheels of the body, an open wheel car can be placed in front of the wing to analyse and gain greater understanding.

References

1. S. Durrer, Aerodynamics of Race Car Wings: A CFD Study, Master's Thesis, Grand Valley State University. http://scholarworks.gvsu.edu/theses/798
2. M.D. Soso, P.A. Wilson, *The influence of an upstream diffuser on a downstream wing in ground effect*. 551–563. https://doi.org/10.1243/09544070JAUTO284
3. W. Mokhtar, S. Durrer, *A CFD Analysis of a Race Car Front Wing in Ground Effect* 2016 ASEE North Central Section Conference
4. S.C. Kachare, *A CFD Study of a Multi-Element Front Wing for a Formula One Racing Car*. https://scholarworks.gvsu.edu/cgi/viewcontent.cgi?article=1864&context=theses

5. A.E. Senior, X. Zhang, *The force and pressure of a diffuser-equipped bluff body in ground effect.* J. Fluids Eng. **123**(1), 105–111 (2001). https://doi.org/10.1115/1.1340637
6. M.D. Soso, P.A. Wilson, *Aerodynamics of a wing in ground effect in generic racing car wake flows.* Proceedings of the Institution of Mechanical Engineers, Part D: Journal of Automobile Engineering, 1–13. https://doi.org/10.1243/2F095440705X69632
7. E. Andrea, X. Zhang, *Influence of diffuser angle on a bluff body in ground effect.* J. Fluids Eng. **125**(2): 332–338 (2003). https://doi.org/10.1115/1.1537252
8. N. Moore, P.A. Wilson, A.J. Peters, 2002 *Aerodynamics of a wing in ground effect in generic racing car wake flows.* Proceedings of the Institution of Mechanical Engineers, Part D: Journal of Automobile Engineering, Volume 220 Issue 1, January 2006. https://doi.org/10.1243/2F095440705X69632
9. *K-epsilon turbulence model.* https://en.wikipedia.org/wiki/K-epsilon_turbulence_model

A Review on Design Optimization of Leaf Spring

Ranjeet Kumar Singh and Vikas Rastogi

1 Introduction

In recent years, the automobile industries have focused on weight reduction of vehicles without any alteration in its payload capacity. Vibrational characteristics, ride comfort, braking performance, and fatigue life of the vehicle have influenced by unsprung weight of the vehicle. Small reduction in this weight improves ride comfort and fatigue life. Leaf spring contributes (1/5) wt% of unsprung weight of the vehicle. Thus, a lot of scope is in reduction of weight in leaf spring. This is achieved by the selection of materials having high strength to weight ratio and proper application of design optimization tool. Development of new materials like functionally graded materials and modern optimization tools will play important role in the fabrication of leaf spring.

Optimization is defined as the process of setting up the various parameters by satisfying the various constraints involved and in case of leaf springs used in automobile it is about spring weight, spring length, thickness, width, load-carrying capacity, and many more parameters. For design optimization of leaf springs Genetic Algorithm, ANSYS, Abacus, NX NASTRAN, Noesis Optimus have used by different authors.

In this paper, a detailed literature survey on design and optimization of leaf spring has been performed. The main objective of this study is to provide the most suitable material and optimization technique for leaf spring. A lot of design and optimization techniques are adopted by the researchers for conventional and fiber composite leaf springs but still there is the scope of introducing new materials and new techniques like functionally graded materials and Python. This literature survey has revealed that a lot of authors have adopted ANSYS for the design and optimization of leaf

R. K. Singh (✉) · V. Rastogi
Department of Mechanical Engineering, Delhi Technological University, Delhi 110042, India

R. K. Singh
Department of Mechanical Engineering, G.L. Bajaj Institute of Technology and Management, Greater Noida, India

B. Popuri et al. (eds.), *Recent Trends in Engineering Design*, Lecture Notes in Mechanical Engineering, https://doi.org/10.1007/978-981-16-2900-6_9

spring, whereas some authors adopted genetic algorithm well as few authors adopted NX NASTRAN, Noesis Optimus, Abacus.

2 Literature Review

More authors have adopted ANSYS for the design and optimization of leaf spring. Shokrieh et al. [2] have shown their interest in optimizing using ANSYS V5.4 software and succeeded to reduce spring mass by 80% in absence of eye units and decrement in width and increment in thickness. Bhandarkar and Shekhawat [3] worked on designing, optimizing, and analyzing using FEM analysis and observed this FEM methodology as the best method in order to get ride comfort, increased life and optimization and verification in less time. Shankar and Vijayarangan [4] have fabricated composite mono leaf spring for light vehicles made by GFRP, graphite/Epoxy, and carbon/Epoxy. Design and optimization of the leaf spring has performed by using ANSYS. Patunkar and Dolas [5] have used ANSYS 10 for computational analysis of composite leaf spring. Considering spring width and thickness are design variables, it is found that 84.40% weight reduced under same payload conditions. Kumar et al. [6] have presented an approach toward optimization of leaf spring. ANSYS and FEM are used for various composite material. E-Glass/Epoxy reduces weight by 85%. 94.18% by graphite eproxy. 92.94% by carbon epoxy. Strain Energy, spring weight, and thickness spring are considered for optimization. Ghodake and Patil [7] have used GFRP and the polyester resin (NETPOL 1011) for fabrication of leaf spring. Optimization and design has been performed by using ANSYS. Fabricated leaf is 84.94% lighter than the conventional. Saini et al. [8] have used ANSYS for optimization for composite material leaf spring. Observation revealed that leaf spring made by composites material has weight reduction of 81.22% over conventional leaf springs. Yadav et al. [9] have selected 55Si2Mn90, 38Si6 material to design leaf spring. Observation has reviled that the deflection leaf spring made by 55Si2Mn90 is f 0.73011 mm which is less than the spring made by 38Si6. 55Si2Mn90 leaf spring has good performance. Kumar et al. [10] worked on design optimization of heavy vehicle leaf spring. Investigation has showed that by replacing the conventional spring material with S2 glass composites weight is reduced from 267kgs to 264kgs. Nikhil et al. [11] have replaced conventional steel leaf spring with mono-parabolic composite leaf spring. Design and optimization have done by using ANSYS. Composite leaf reduces the stress up to 30% and saves up to 80% of the entire weight as compared to the conventional steel material. Natalupati [12] has used fiber reinforced plastics to fabricate mono composite leaf spring. The leaf spring model was done by using Pro/ENGINEER and design analysis performed with the help of ANSYSIS12.0. Observation has revealed that 85% of conventional leaf spring is saved by using GFRP leaf spring. Kaveri et al. [13] have applied ANSYS and CATIA for design and analysis of GFRP leaf. It is observed that reduction in weight of composites for E-glass/Epoxy leaf spring is nearly 70%. Kumar and Aggarwal [14] used ANSYS and selected steel as leaf spring material and observed that reduction of 1.16% of stress in parabolic design is due to elimination of interleaf friction. The fatigue life is 4.83% more in parabolic design as compared to conventional leaf spring. The weight of parabolic design is decreased by 9.32% in comparison to conventional multi-leaf spring. Krishan and Aggarwal [15] have applied ANSYS and used GFRP composite for the fabrication of leaf spring. Analysis has showed that the alternating stress remains in the limits, only a variation of 2.66% is noticed. The fatigue life is 3.73% more in GFRP leaf spring as compared to conventional steel parabolic leaf spring. The weight of GFRP leaf spring is decreased by 67.72% than steel parabolic leaf spring. Madhava and Deepak [16] have worked on design and optimization of leaf spring by using ANSYS 14.5. Leaf spring was fabricated by grey cast iron, plain carbon steel. Manjit et al. [17] have worked on design optimization through ANSYS 14.5 and presented Grey relation analysis and observed minimum stress and deflection simultaneously drawn on Carbon Epoxy material. Vijay and Munde [18] used ANSYS and observed weight reduction of 19.39% without any compromising strength and factor of safety. Solanki and Kaviti [19] used ANSYS 14.0 and concluded that parabolic leaf spring shows better result as compared to semi-elliptical at same conditions and applied load. By decreasing the span length, the deflection and stress for semi-elliptical is 5.081 and 23.584, respectively, which is higher as compared to the parabolic leaf whose deflection and stress is 3.7989 and 21.901, respectively. Teli et al. [20] have used EN45, GFRP materials for the fabrication of leaf spring. In optimization process, deflection, stiffness, energy absorbed, and natural frequency are considered as design parameters. It is found that weight reduction is by 67.70% of composite material by using ANSYS. Dr. Barnabas and Kingston [21] presented optimization work using Taguchi's DOE. Investigation has showed the lowest value of deformation obtained as 19.919 mm which was lowered by 8.122%. Nayak et al. [22] worked on leaf spring optimization and noticed soring with composites resulted in 40% weight reduction. Bathuka et al. [23] used ANSYS and observed deflection for various materials like Carbon Epoxy and S2 Glass showed much better results as compared to mild steel also stress and strain generated were same for all. Pradip and Matawale [24] used ANSYS 15 as an optimization tool, and considered 50Cr1V23, 50Cr1, 38Si6, 55Si2Mn90 for the optimization process and concluded that among all the material used for making steel leaf spring 55Si2Mn90 was the most optimum one. There was a 3.3mpa reduction in stress for 55Si2Mn90 as compared to 50Cr1V23 and less deflection in 55Si2Mn90 as compared to among all the materials.

Some authors have applied genetic algorithms for optimization of leaf spring. Rajendran et al. [1] have used genetic algorithm for the optimization of conventional and glass fiber composite leaf spring. GA includes various design constraints and selection procedures. The fittest over the all is selected for designing. Observation has showed 8% reduction of steel spring weight using genetic algorithm. Kassie et al. [25] have performed design optimization on mono-composite leaf spring. Optimization was done by using genetic algorithm. During optimization weight is chosen as objective function. It is observed that 68.14% of weight reduction with same load- carrying capacity and strength. Qian et al. [26] have used genetic algorithm for the optimization of FRP leaf spring. Thickness is considered as design variables. Observation has showed that 0.58% improvement in Fatigue life and stiffness of the leaf spring after optimization. Li et al. [27] have used Carbon fiber, glass fiber, and Aluminum Alloys for the fabrication of composite leaf spring suitable for electric aircraft. Conduction of optimization on position of fiber with leaf spring axis revealed that ±45° is best-suited position. The weight of composite leaf spring is reduced from 4.8 kg to 3.8 kg after using GA as optimization tool.

Few authors adopted NX NASTRAN, Noesis Optimus, Abacus for the design and optimization of leaf spring. Jayakanth et al. [28] have worked on design analysis of commercial vehicle leaf spring. Leaf spring is fabricated by carbon steel material. Optimization was done by using NX NASTRAN. The weight of fabricated leaf spring is reduced by 40% in comparison to conventional spring. Kadziela et al. [29] have used LMS Virtual. Lab Motion for modeling of suspension provides fast and easy modeling of leaf spring suspension and time reduction for full vehicle dynamic modeling. Optimization of leaf spring has been done by using Noesis Optimus software. Palaskar [30] has used Abacus for design and optimization of mono-leaf spring. Optimization of mono-leaf spring and noticed material saving up to 25% by parabolic variation in thickness along length of leaf spring by using Abacus. Table 1 shows the materials, optimization tools, optimization variables used in the design of leaf spring. Major findings have also showed in tabular form.

3 Result and Discussion

Literature survey has reviled that the progress in the field of design and optimization in previous decades increases rapidly as shown in Fig. 2. Design optimization tools adopted by researchers are ANSYS, ABACUS, Genetic Algorithm, NX NASTRON, and Noesis Optimus. Contribution of each tools for design optimization of leaf spring is shown in Fig. 1. Introduction of modern design and optimization techniques and development of new materials also gives an opportunity for researchers in the development of leaf spring which will primary suspension in future flying vehicles.

4 Conclusion

The exclusive literature survey has been executed on design optimization of leaf spring. Outcomes of this survey concluded as follows:

1. Most of the authors have adopted ANSYS for design optimization of leaf spring; thus there is still scope of other computational techniques like ABACUS, G.A., NASTRAN, Noesis Optimus, etc.
2. In the design optimization of leaf spring, it is found that fiber composites are more suitable candidate in comparison to the conventional steel leaf springs but there is still scope of functionally graded material.

Table 1 Major findings in the design optimization of leaf springs

S. no	Type of material	Materials used	Optimization tools	Optimization variables	Result	Authors
1.	Composite material	Fiber Reinforced Plastic	Genetic Algorithm	Width, Thickness	• Optimization helped in the reduction of 8% of steel leaf spring and 23.4% of composite leaf spring weight • Mono-composite leaf is 75.6% lighter in weight as compared to multi- leaf spring	[1]
2.	Composite material	E- poxy/E- glass	ANSYS V5.4 Software	Spring width, Thickness	• It was observed that spring weight get lowered by 80% in absence of eye units and decrement in width and increment in thickness	[2]
3.	Composite material	GFRP, graphite/Epoxy, and carbon/Epoxy	ANSYS Software	Width and Thickness	• 85%, 91%, and 90% of the weight is reduced by using GFRP, graphite/Epoxy, and carbon/Epoxy composites	[4]
4.	Composite material	GFRP	ANSYS 10.0	Spring width, and thickness	• E-Glass/Epoxy mono-leaf spring indicates weight reduction by almost 84.40% with better performance	[5]

(continued)

Table 1 (continued)

S. no	Type of material	Materials used	Optimization tools	Optimization variables	Result	Authors
5.	Composite material	Carbon/epoxy Graphite/epoxy E-glass/epoxy Fiber reinforced plastic	ANSYS	Thickness & length	• Displacement of leaf spring is 92.5 mm Chamber length of the leaf spring is 180 mm	[6]
6.	Steel and composite material	GFRP with polyester resin (NETPOL 1011)	ANSYS	Spring width and thickness	• It was observed that weight got reduced in case of mono-composite leaf spring by 84.94%	[7]
7.	Steel and composite spring	Steel, plain Carbon (0.90% to 1%), Carbon graphite, and glass fiber	ANSYS 9.0	Full & half-length and radius of curvature	• Reduces weight of the composite leaf spring is 81.22%	[8]
8.	Composite material	Cast Iron	ANSYS VS.4	Thickness and cross-section	• Decrease stress concentration of the composite leaf spring Increase weight of the leaf spring	[3]
9.	Steel material	55Si2Mn90, 38Si6	ANSYS, AutoCAD	Stress and deflection	• Minimum deflection of 0.73011 mm with good performance was observed in case of spring made of 55Si2Mn90	[9]

(continued)

Table 1 (continued)

S. no	Type of material	Materials used	Optimization tools	Optimization variables	Result	Authors
10.	Composite material	Epoxy/E- glass	Genetic Algorithm	Deflection, Longitudinal and transverse tensile strength, Shear strength, Strain	• Composite leaf spring is 68.14% lighter as compared to steel leaf spring • Material is not reliable because of the chipping problem in bumpy road, it has an acceptable fatigue life of 221×10^3 cycles	[25]
11.	Steel	Isotropic material	ABACUS	Parabolic variation of thickness. parabolic, leaf spring, thickness, deflection, stress	• Up to 25% Material saving can be achieved by parabolic variation in thickness along the length of the leaf spring	[30]
12.	Composite material	Epoxy/E-glass	ANSYS 10.0	Leaf Spring, Static Analysis, Dynamic Analysis FEA	• By comparing the results for 3 materials, using Epoxy Matrix Composite reinforced by 50% of Kevlar fibers is better since its • Weight is less	[10]

(continued)

Table 1 (continued)

S. no	Type of material	Materials used	Optimization tools	Optimization variables	Result	Authors
13.	Composite material	Carbon Epoxy, 55Si2Mn90(steel)	ANSYS	Von Mises Stress, Total Maximum Deflection, Total Mass	• By replacing conventional steel leaf spring with mono-parabolic composite leaf spring, and found that stress is reduced by 30% • The composite material accumulates up to 80% of the entire weight as compared to the conventional steel material • The fatigue life of mono-parabolic leaf spring is higher than the conventional steel leaf spring	[11]
14.	Composite material	Fiber Reinforced Plastic (Epoxy/E-glass)	ANSYS	Stresses and deformation	• Reduction of weight by 84% was observed when convention material of spring was replaced by composite mono-leaf spring	[12]
15.	Composite material	Glass Fiber Reinforced plastic	ANSYS 12.1, CATIA	Ultimate Compressive strength, Ultimate Tensile strength	• The reduction in weight of composite for E-glass/Epoxy leaf spring is nearly 70%	[13]

(continued)

Table 1 (continued)

S. no	Type of material	Materials used	Optimization tools	Optimization variables	Result	Authors
16.	Steel and composite material	Steel, Fiber reinforced plastics	Noesis Optimus	Weight and crossover rate	• Friction coefficient(μ = 0.1) was increased and Frequency was found to be usually 1 Hz	[29]
17.	Steel material	EN45A	ANSYS	Von Mises Stress, Fatigue life, Weight, Displacement	• Reduction of 1.16% of stress developed in parabolic design due to elimination of interleaf friction • The fatigue life is 4.83% more in parabolic design as compared to conventional leaf spring. • The weight of parabolic design is decreased by 9.32% in comparison to conventional multi leaf spring	[14]
18.	Composite material	Glass Reinforced Plastics	ANSYS	Alternating stress, Fatigue life, Weight	• The fatigue life is 3.73% more in GRP leaf spring as compared to steel parabolic desig • The weight of GRP leaf spring is decreased by 67.72% which makes it lighter than steel parabolic leaf spring	[15]

(continued)

Table 1 (continued)

S. no	Type of material	Materials used	Optimization tools	Optimization variables	Result	Authors
19.	Composite and steel material	Grey cast iron& plain carbon steel	ANSYS 14.5, SOLIDWORKS	Optimized stress &strain(length, width)	• Improved damping capacity and Less deformation	[16]
20.	Composite material	AISI1008 Carbon Steel	NX Nastran	Spring length, spring width and stresses.	• With respect to conventional leaf spring, 40% of spring weight and un sprung mass was reduced with composite AISI1008.	[28]
21.	Composite material	Fiber Reinforced plastics	Genetic Algorithm	Length and width	• Improve Fatigue life of the leaf spring Stiffness before and after optimization was 0.58%	[26]
22.	Composite material	Carbon E- poxy	ANSYS 14.5	weight savings	• Carbon Epoxy material have minimum stress and deflection	[17]
23.	Composite material	Fiber Reinforced plastics	ANSYS 16.0 and topology optimization	Total Deformation, Equivalent Stress, Safety Factor	• In this case, weight reduction of 19.39% is observed almost without compromising strength and factor of safety	[18]

(continued)

Table 1 (continued)

S. no	Type of material	Materials used	Optimization tools	Optimization variables	Result	Authors
24.	Steel material	Stainless Steel	ANSYS 14.0	Deflection and Stress	• Parabolic leaf spring performs better as compared to semi-elliptical under same conditions and load application • The deflection and stress for semi-elliptical is higher than the parabolic leaf	[19]
25.	Steel and composite material	EN45, Glass Reinforced plastics	ANSYS	Deflection, Stiffness, Energy absorbed, Natural Frequency	• Weight reduction by 67.70% • Fatigue analysis estimates the life of the leaf spring	[20]
26.	Steel and composite Material	Carbon Epoxy, Graphite Epoxy, Conventional steel	ANSYS 18.0 and Taguchi's DOE	Number of leaves and length of leaves thickness	• The lowest value of deformation obtained in the confirmation experiment with the use of Taguchi 's DOE is 19.919 mm • which is lowered by a least of 8.122% than all the obtained experimental values	[21]

(continued)

Table 1 (continued)

S. no	Type of material	Materials used	Optimization tools	Optimization variables	Result	Authors
27.	Composite material	C-glass fiber E-glass/Epoxy Silicon spray	ANSYS	Three point load in the middle of the leaf spring.	• 40% reduction of weight of composite leaf spring • Stress concentration is reduced 76.39% error percentage is 0.84% for the numerical values	[22]
28.	Composite and steel material	Mild Steel, Carbon Epoxy, S2 Glass	ANSYS R14.5	Deflection, Stress, Strain, Fatigue life	• Carbon Epoxy and S2 Glass showed much better results as compared to mild steel. • Stress and strain generated were same for all	[23]
29.	Composite material	Carbon fiber, glass fiber, Aluminum Alloys	Genetic Algorithms	Thickness of layup	• Best result achieved when composite fibers at an angle $\pm 45°$. • Weight of composite leaf is 4.8 kg before optimization and 3.8 kg after optimization	[27]
30.	Steel material	50Cr1V23, 50Cr1, 38Si6, 55Si2Mn90	CATIA V5 R20, ANSYS 15	Von Mises Stress, Deflection	• Steel leaf spring made by 55Si2Mn90 has optimum weight • There was a 3.3 MPa reduction in stress for 55Si2Mn90 as compared to 50Cr1V23. • Lowest deflection in 55Si2Mn90	[24]

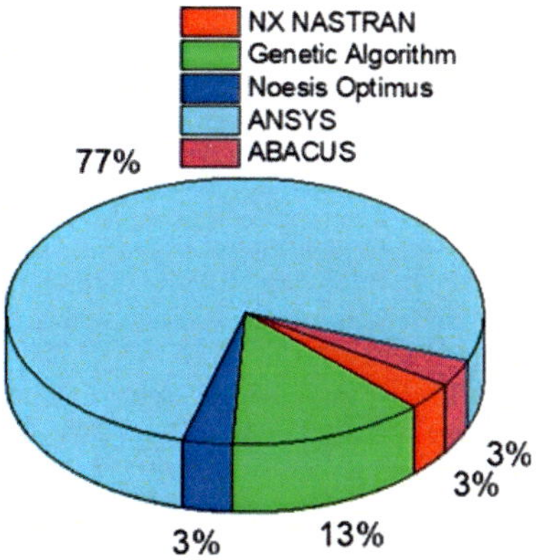

Fig. 1 Percentage of design optimization tools used by researchers

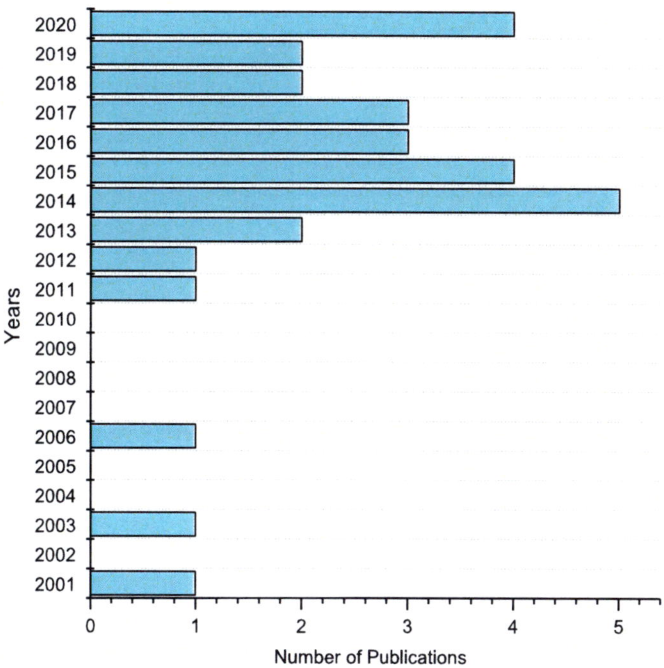

Fig. 2 Yearwise publications on design optimization of leaf springs

3. In multi-objective optimization, results obtained by using G.A. are good thus Genetic Algorithm is most suitable tool.
4. During optimization, most of the researchers followed width, thickness, deflection, and stresses of leaf springs as design variables as well as weight was considered as objective function.

References

1. I. Rajendran, S. Vijayarangan, Optimal design of a composite leaf spring using genetic algorithms. Comput. Struct. **79**(11), 1121–29 (2001). https://doi.org/10.1016/S0045-7949(00)00174-7
2. M.M. Shokrieh, D. Rezaei, Analysis and optimization of a composite leaf spring. Compos. Struct. **60**(3), 317–325 (2003). https://doi.org/10.1016/S0263-8223(02)00349-5
3. D.K. Bhandarkar, S.P. Shekhawat, Design, analysis and optimization of a leaf spring. Int. J. Innov. Res. Sci. Technol. **3**(6), 13658–13666 (2014)
4. G. Shankar, S. Vijayarangan, Mono composite leaf spring for light weight vehicle-design, end joint analysis and testing. Mater. Sci. **12**(3), 220–225 (2006)
5. M.M. Patunkar, D.R. Dolas, Modelling and analysis of composite leaf spring under the static load condition by using FEA. Int. J. Mech. Industr. Eng. **1**(1) (2011)
6. M.A. Kumar, T.N. Charyulu, Ch. Ramesh, Design optimization of leaf spring. Int. J. Eng. Res. Appl. (IJERA) **2**(6), 759–765 (2012)
7. A.P. Ghodake, K.N. Patil, Analysis of steel and composite leaf spring for vehicle. IOSR J. Mech. Civil Eng. **5**(4), 68–76 (2013). https://doi.org/10.9790/1684-0546876
8. P. Saini, A. Goel, D. Kumar, Design and analysis of composite leaf spring for light vehicles. Int. J. Innov. Res. Sci. Eng. Technol. **2**(5) (2013)
9. N. Yadav, P. Tandon, S.A.K. Jilani, Material optimization of leaf spring of tractor trolley by FEA. Ijmer **4**(4), 40–44 (2014). http://www.ijmer.com/papers/Vol4_Issue4/Version-2/IJMER-44024044.pdf
10. T.N.V.A. Kumar, E.V. Rao, S.V.G. Krishna, Design and material optimization of heavy vehicle leaf spring. Int. J. Res. Mech. Eng. Technol. **5762**(1), 80–88 (2014)
11. Karlus, E. Nikhil, R.L. Himte, R.K. Rathore, Optimization of mono parabolic leaf spring **7**(1), 283–91 (2014)
12. S. Nutalapati, Design and analysis of leaf spring by using composite material for light vehicles. Int. J. Mech. Engi. Technol. **6**(12), 36–59 (2015)
13. K.A. Kaveri, Mankar, S.H., D.J. Samir, A review on design and optimization of composite leaf spring. Int. J. Innov. Emerg. Res. Eng. **2**(1), MEPCON (2015)
14. K. Kumar, M.L. Aggarwal, Optimization of various design parameters for EN45A flat leaf spring. Mater. Today: Proc. **4**(2), 1829–1836 (2017). https://doi.org/10.1016/j.matpr.2017.02.026
15. K. Krishan, M.L. Aggarwal, Simulation for optimized modelling of En45A leaf spring. Int. J. Recent Adv. Mech. Eng. **4**(3), 129–42 (2015). https://doi.org/10.14810/ijmech.2015.4310
16. M. Madhava, G. Deepak, Design and analysis of functionally graded leaf spring structure. Int. J. Eng. Res. **5**(6), 1129–1254. https://doi.org/10.17950/ijer/v5i6/023
17. L. Manjit, M. Sahu, A.K. Khandelwal, S.P. Shrivas, 092017 analysis and optimization of leaf spring **6**(3), 429–34 (2017)
18. K. Vijay, K.H. Munde, A. Pawar, Topology optimization of front leaf spring mounting bracket **3**(7), 12–19 (2018)
19. P. Solanki, A.K. Kaviti, Design and computational analysis of semi-elliptical and parabolic leaf spring. Mater. Today: Proc. **5**(9), 19441–19455 (2018). https://doi.org/10.1016/j.matpr.2018.06.305
20. Teli, D. Mayur, U.S. Chavan, H.G. Phakatkar, Design, analysis and experimental testing of composite leaf spring for application in electric vehicle. Int. J. Innov. Technol. Exploring Eng. **8**(9), 2882–91 (2019). https://doi.org/10.35940/ijitee.i8744.078919
21. Barnabas, J. Kingston. Optimization of leaf spring parameters using taguchi 's DoE **8**(12), 6–18 (2019)
22. S. Nayak, J. Sadarang, I. Panigrahi, R.K. Nayak, M. Maurya, Optimization of composite leaf spring for reduced weight and improved noise, vibration, and harshness in an electric vehicle. Noise Vibr. Worldwide **51**(7–9), 127–138 (2020). https://doi.org/10.1177/0957456520923319

23. M. Bathuka, B. Gupta, S. Kranthi Kumar, S.P. Jani, Modeling and analysis of leaf spring with different type of materials. Mater. Today: Proc. (2020). https://doi.org/10.1016/j.matpr.2020.09.223
24. K. Pradip, C.R. Matawale. Analysis and optimization of mono parabolic leaf spring material using ANSYS. Mater. Today: Proc. **33**, 5757–64 (2020). https://doi.org/10.1016/j.matpr.2020.06.605
25. A. Kassie, Achamyeleh, Design of single composite leaf spring for light weight vehicle. Int. J. Mech. Eng. Robot. Res., 34–41 (2014). https://doi.org/10.23883/ijrter.2017.3192.jtt1n
26. C. Qian, W. Shi, Z. Chen, S. Yang, Q. Song, Fatigue reliability design of composite leaf springs based on ply scheme optimization. Compos. Struct. **168**, 40–46 (2017). https://doi.org/10.1016/j.compstruct.2017.02.035
27. D. Li, Z. Fan, Y. Zhang, J. Zang, F. Yang, Optimum design and experiment of composite leaf spring landing gear for electric aircraft. Chin. J. Aeronaut. **33**(10), 2649–2659 (2020). https://doi.org/10.1016/j.cja.2020.03.010
28. J.J. Jayakanth, M. Jeyaraman, S. Sivaganesan, M. Chandrasekaran, Design and analysis of commercial vehicle leaf spring using aisi1008 carbon steel composite material. ARPN J. Eng. Appl. Sci. **11**(17), 10387–10391 (2016)
29. B. Kadziela, M. Manka, T. Uhl, A. Toso, Validation and optimization of the leaf spring multi-body numerical model. Arch. Appl. Mech. **85**(12), 1899–1914 (2015). https://doi.org/10.1007/s00419-015-1024-5
30. M.P. Palaskar, Optimization of mono leaf spring by parabolic variation of thickness along the length **4**(5), 1–11 (2014)

Numerical Study of Twisted Tape with Circular Cutout and Triangular Cutout in a Circular Tube

Awasthi Aditya Bachchan, Sourabh Gahlot, Gopal Nandan, Satish Kumar, and Ramakant Shrivastava

1 Introduction

Enhancement of heat transfer in heat exchanging devices using various techniques has been attempted and utilized in many engineering applications. Usage of twisted tube as turbulence promoters is a passive technique. It is being widely served for improving the thermal performance in the heat exchangers. Heat exchangers assist the heat transfer in between two or more fluids. The heat transfer may take place from hot to cold fluid and vice versa. Fluids in heat exchangers may be in direct contact with each other or may be separated by a boundary. Constructions like cooling towers, desuperheaters, jet condensers, open feed water heaters, etc., are the types of direct contact type heat exchangers, whereas the preheaters used in blast furnaces, oxygen plants, etc., are forms of regenerators [1]. They are used in every industry whether it is a manufacturing plant or a power plant, an automobile or a refrigerator. For instance, a car radiator is a heat exchanger that cools down the hot water from the engine and transfers heat to the atmospheric air. Some more examples of heat exchangers under action are waste heat recovery, space heating, chemical processing, air conditioning, etc. [2]. A heat exchanger with high performance is very much essential as it helps in saving energy [3]. They are vital in a variety of fields such as refrigeration, metallurgy, chemical engineering [4]. They come in various shapes and sizes. Both large and small heat exchangers have their pros and cons. In a small heat exchanger, rate of heat exchange is low and in large heat exchangers, the mere large size that acts as a backdrop [5]. In this present time, the designers venture to design

A. A. Bachchan
Amity University Uttar Pradesh, Noida, India

S. Gahlot · S. Kumar
National Institute of Technology, Jamshedpur, India

G. Nandan (✉) · R. Shrivastava
Government College of Engineering, Karad, India
e-mail: gnandan@amity.edu

B. Popuri et al. (eds.), *Recent Trends in Engineering Design*, Lecture Notes in Mechanical Engineering, https://doi.org/10.1007/978-981-16-2900-6_10

compact heat exchangers with greater heat transfer capacity [6]. The heat transfer rate depends upon numerous parameters like pressure drop, thermal resistance between the fluids, turbulence, etc. At the solid–liquid interface, the heat is transferred by convection and conduction in the wall which comes between two fluids. To keep an account of contribution of all these effects, keeping note of overall heat transfer coefficient is conducive [7]. Various arrangements have been used to decrease the thermal resistance of the fluids by increasing the turbulence [8]. In the pursuit of improving the heat transfer capacity, various accessories have been developed, one such arrangement is the application of twisted tape in the flowing fluids [9]. Twisted tape heat exchangers are the passive turbulators that are used as an enhancement in heat exchangers [10]. There are various types of augmentation techniques used along with the twisted tape turbulator that is conducive to increasing the turbulence [11]. This helps in heat transfer enhancement by increasing the turbulence in the flowing fluid. The fluid flowing adjacent to the tube surface receives more heat than the fluid flowing in at a certain distance from the surface of the tube. The temperature of the fluid particles adjacent to the tube surface becomes equal to that of the tube surface and the temperature decreases till we reach the centre. This temperature difference causes initiation of convection and develops a Thermal Boundary Layer (TBL). The thickness of TBL increases along the direction of the flow. At the centre of the tube, the fluid temperature is same as that at the entry of the tube. Using a twisted tape breaks this thermal boundary layer and causes total intermixing of the fluid particles at the centre of the tube and the particles adjacent to the tube by generating swirl flow [12]. Thus, the temperature of the whole fluid irrespective of its position in the tube increases. It reduces the thickness of stagnant fluid sub-layer near to solid–liquid interface. It generates instability in the path of the flow and ultimately increases heat transfer. Labib et al. [13] experimentally studied the heat transfer enhancement due to convection in counter-flow double tube heat exchanger. The data collected from the experiment was compared to a heat exchanger of similar dimension. The twisted tapes used were 4.5–7.5 in twist ratio and the Re ranged in between 20000 and 50000. Significant increase in factors indicating the performance of the twisted tape was observed. However, with decrease in twist ratio the performance indicating factors or the thermal performance factors surged. This surge in performance was limited to specific mass flow rate after which the variation ceased to occur. Researchers have attempted several other types of twisted tape with water, oil, air and nanofluids as working fluids. Bas and Ozceyhan [14] investigated the effect of twisted tape in the heat exchanger. With increase in the heat transfer rate, there will be increase in the friction factor due to obstacle offered by the twisted tape. Hence, simultaneous effect of heat transfer rate as well as study of pressure drop must be studied. A separation is kept in between the tape and the tube wall. The range of Reynolds number is kept in between 5132 and 24989. They also considered other parameters like twist ratio, clearance ratio, etc. Twisted tape enhances the heat transfer rate and pressure drop considerably. With increase in Nusselt number and Reynolds number, clearance ratio and twist ratio decrease. It gives the best performance at lower Reynolds number. Hence, at lower Re, the size of the heat exchanger will be decreased for the same thermal performance. Promevonge et al. [15] studied the effect of a combined twisted tape and conical ring insert in a tube. The external wall of the heat tube was heated electrically along the axis with fixed heat flux boundary conditions. Using air is as the working fluid, all experiments were conducted in the turbulent flow regimes (6000 < Reynolds number <26000). In each run, two twisted tapes were introduced with Y = 3.75 and 7.5. It was revealed from the results that when both the attachments were used Nusselt number value increased from4 to 10% and the efficiency from 4 to 8% than when only conical attachment was used. The maximum heat transfer enhancement of 367% was achieved. It was also found that for Reynolds number greater than 16000 the enhancement efficiency remained constant and with further increase in Reynolds number it decreases.

Abed et al. [16] performed numerical investigation in a horizontal tube to study the effect of V-cut twisted tapes and perforated twisted tapes in the heat transfer characteristics and the flow characteristics. The numerical simulation of water flowing in horizontal tube with or without twisted tape insert is performed. The horizontal tube is subjected to uniform heat flux. Realizable K-ε model is used for the simulation and the solution is done using finite volume method. The Reynolds number ranged from 40000 to 9000 and the heat flux subjected on the tube ranged from 5000 to 1000 W/m^2. V-cut and Perforated twisted tapes of twist ratio 4–6 were considered in this numerical simulation. Comparison between pipe with twisted tapes and one without twisted tape was done on the basis of thermal performance and average and local Nusselt number. The effect of boundary conditions such as Reynolds number and twist ratio on the parameters was also studied. It was observed from the results that friction factor and Nusselt number were proportional to the twist ratio for all Reynolds number. The thermal performance factor was observed to be directly proportional to the Reynolds number but inversely proportional to the twist ratio of the tape. The V-cut twisted tape of twist ratio 4 showed greater performance than the perforated twisted tape of twist ratio 4.

Salam et al. [17] experimentally investigated the heat transfer augmentation in tubular heat exchangers using twisted tapes with rectangular cutout. The investigation was carried out in order to measure the parameters such as heat transfer enhancement, friction factor and coefficient of heat transfer for inside the tube. The dimensions of the copper tube used in the experiment are internal diameter of 26.6 mm, external diameter of 30 mm and 900 mm in length, and the dimensions of the cutout section are width of 14 mm and depth of 8 mm. To ensure uniform heat flux over the section under observation, fibre glass and Nichrome wire were wrapped over it. Thermocouples were used to measure the temperature at the outer surface of the tube and to measure the bulk temperature two thermometers were used which were placed in the mixing box at the outlet. The range of Reynolds number was 10000–19000 and for heat flux 14–22 kW/m^2 and 23–40 kW/m^2 for smooth tubes and tubes with inserts, respectively. It was concluded from the results that the Nusselt number is proportional to the Reynolds number. Using the inserts with rectangular cutouts helped in achieving 68% more heat transfer.

Thianpong et al. [18] investigated the pressure drop and heat transfer characteristics of flow inside a heated tube with perforated and winged twisted tapes. The experiment was conducted for flow in turbulent range, i.e. for Reynolds number in range of 5500–20500. The concept of wings was introduced in the twisted tapes as it increases the turbulence and yields disturbance in thermal boundary layer and the perforation along the tape causes less pressure drop. The dimensions of the twisted tape are hole diameter ratio of 0.11–0.55 and width ratio of 0.11–0.33 for the perforation and the wings, respectively. It was concluded from the result that the perforated and winged twisted tape helped in achieving 208% enhancement in heat transfer whereas plain twisted tape augmented heat transfer by 190% only, when compared to a smooth pipe. The maximum thermal performance factor of 1.32 was achieved by using perforated and winged twisted tape of 0.11 hole diameter ratio and wing depth ratio of 0.33 at Reynolds number of 5500.

Farnam et al. [19] studied the enhancement in heat transfer in an agitated U tube due to twisted tube and twisted tapes, with the help of experimental investigation. Two different twisted tapes were employed in order to study the parameters affected by the twist ratio. A comparison between the results of plain tube and twisted tube was done for Reynolds number in range of 4250–12750. It was concluded from the results that straight tube and twisted tube with twisted tapes showed greater heat transfer coefficient than those without twisted tapes. It was also observed that both heat transfer coefficient and pressure drop show inverse relation with the twist ratio. The friction factor and Nusselt number achieved by twisted tube were greater than that achieved from plain tube by 335 and 58%, respectively. The twisted tube with twisted tapes attached yielded 46–82% and 33–168% higher friction factor and Nusselt number, respectively. The twisted tube with twisted tape of lower twist ratio yielded a maximum performance index of 2.28.

2 Literature Review of Water as Working Fluid

The effect of twisted tape in the combination of alternating clockwise and anti-clockwise was studied in a tube type heat exchanger [20]. Construction of nine twisted tapes was done with twist ratios 3.0–5.0 and angle of twists 30° to 60°, and Reynolds number in between 3000 and 27000. The heat transfer rate, heat transfer augmentation and friction factor of simple twisted tape are less than that produced by anti-clockwise twisted tape. With the increase in twist angle value and decreasing twist ratio, the heat transfer rate appears to increase. The value of the average Nusselt number (Nu) according to the twist ratio, Reynolds number and twist angle value, is greater in counter-clockwise twisted tape than in simply twisted tapes and typically twisted tapes by a factor of 27.3–90.5% and 12.8–41.9%. Kumar et al. [21] investigated the friction and heat transfer characteristics of a perforated twisted tape tubular heat exchanger with V-cut. The Reynolds number was in the range of 2700–23400 and the twist ratio varied between 2 and 6. It was concluded from the result that the maximum thermohydraulic parameter is 1.58. Shabanian et al. [22] studied the effect of classic, butterfly and jagged type inserts. The relative thermal performance for butterfly insert was the best. The difference between the heat transfer of jagged and classic inserts decreases by decreasing the twist ratio. They also showed similar observation from analysis using computational fluids dynamics software. Numerical study have been reported using perforated twisted tape for its thermal hydraulic performance. The perforated twisted tapes were developed to eradicate the friction loss in the fluid flow. Twisted tapes of three different diameters and three different pitches were studied with Reynolds numbers in the range of 6000–20000. It was an observation in the result that there was lesser friction loss in perforated twisted tape than conventional twisted tapes. With the decrease in diameter ratio and pitch ratio, there was an increase in the friction loss, heat transfer and CTP. The maximum thermal performance factor achieved was 1.28 at Reynolds number 6000. The twisted tape cut with peripheral cut with water as working fluids were experimentally studied [23]. The twist ratio of 3 was maintained and three different tape depth ratio and tape width ratio was tested. Uniform heat flux was maintained with the isothermal condition and Reynolds number was kept between1000 and 20000. There was a significant development in the heat transfer rate and friction factor in the twisted tape with the peripheral cut. It also helped in achieving greater turbulence intensity. The heat transfer enhancement is proportional to depth ratio and width ratio. The maximum performance factor of 1.28 was achieved. Zhang et al. [24] studied the performance of Self-Rotating Twisted Tapes (SRTT). A comparison was also made between the performance of simple twisted tape and SRTT with twist ratios 2.2, 3, 4, and 6 and Reynolds number in between 12000 and 45000. The SRTT provides a significant increase in the heat transfer performance than that of the simple twisted tapes. They have suggested a correlation between the friction factor and Nu. Mashoofi et al. [25] investigated the ways to decrease the pressure drop and enhance the thermal performance enhancement factor. The investigation was made on axially perforated twisted tapes with holes of varying diameter. The variation in the efficiency with several transfer units for various heat capacity ratios was also exhibited. The validation of the numerical model was also done with the experimental results available. With the help of the results, it was concluded that the perforated twisted tape helps in decreasing the heat transfer rate.

Pourahmad and Pesteei [26] performed the experimental investigation on the thermohydraulic performance of tubes with Reynolds numbers between 4000 and10000. Gnielienski and Filonenko equations were used to determine the Nusselt number and friction factor with an average deviation of 9.5 and 2.8%. In the case of successive twisted tapes tube, the gap between the wall and the inserts is predicted and the deviation of Nu and friction factor lies within 10%. Twisted tapes with a twist ratio of 4 help in obtaining a Coefficient of Thermal Performance (CTP) of 1.08 to 1.10. When raising the gap from 0.043 to 0.135 Nusselt number and friction factor drops by 6.7% and 11.6%, respectively. There is a significant decrease in the CTP and in turbulent transition region, it is reversed negatively. A similar improvement can be obtained by using U cut and isosceles V-cut on the Nu and friction factor. However, by using obtuse V-cut highest CTP of 1.18 to 1.23 was obtained and with a significant increase in Nusselt number and friction factor too.

Naphon [27] conducted experiments in double horizontal pipe. The testing of two different sections was done with different relative pitches. For the inner tube, the inner and outer diameters were set to 8.10 and 9.54 mm. The tapes were 2000 mm in length and 1 mm in thickness. Hot and cold water were used as working fluid in the tube side and shell side. The mass flow rate for cold water was maintained between 0.01 kg/s and 0.07 kg/s and for hot fluid was varied from 0.04 to 0.08 kg/s. The inlet temperature of the cold water was between 15 and 20 °C, whereas the inlet temperature of hot water ranged between 40 and 45 °C. The performance of the twisted tube with twisted tape was examined. Non-isothermal correlations were proposed from the data gathered for friction factor as well as heat transfer coefficient for the horizontal tubes. The resultant data lied within 15–20% of the data yielded from the projected correlation. Pourahmad [26] studied the effectiveness of a wavy strip double tube heat exchanger. Reynolds number for the different flow rates of cold water were studied. The wavy strip turbulator was made of galvanized steel with different angles (45°, 60°, 90°, 120° and 150°) with a thickness 1 mm. The Re was kept between 3000 and 13500. During the experiment, the water was used as hot and cold fluid which were maintained at a constant temperature. It was concluded that there was a noteworthy effect of the turbulator on the effectiveness and NTU.

3 Physical Modelling

Modelling steps have been precisely mentioned in the succeeding section. The inner diameter of the tube is 30 mm and the length is 1 mm. The twist ratio is set as 10 and the width as 24 mm. In the numerical analysis, the thickness is considered negligible and the thickness is 1 mm. For TCTT, the tube is 30 mm internal diameter, the depth of cutout is set as 5 mm. The twist ratio of the twisted tape is set as 10 and the width as 24 mm. In the numerical simulation, the wall thickness is considered negligible and the thickness of the twisted tape is 1 mm. For CCTT, the internal diameter is 30 mm, the depth of cutout is set as 5 mm. The twist ratio is set as 10 and the width as 24 mm. In the numerical analysis, the thickness is considered negligible and the thickness is 1 mm (Figs. 1, 2 and 3).

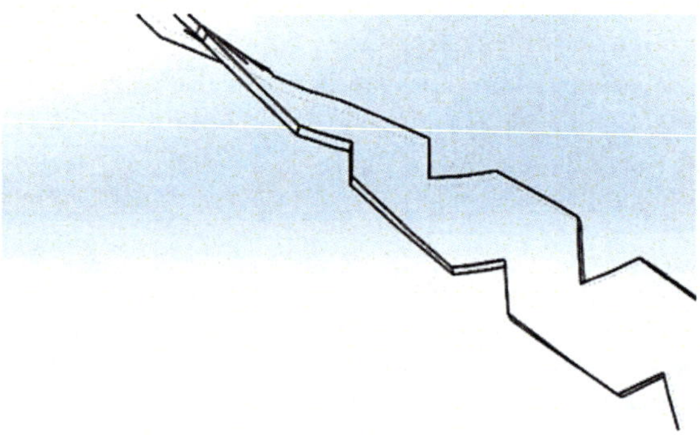

Fig. 1 Triangular cutout

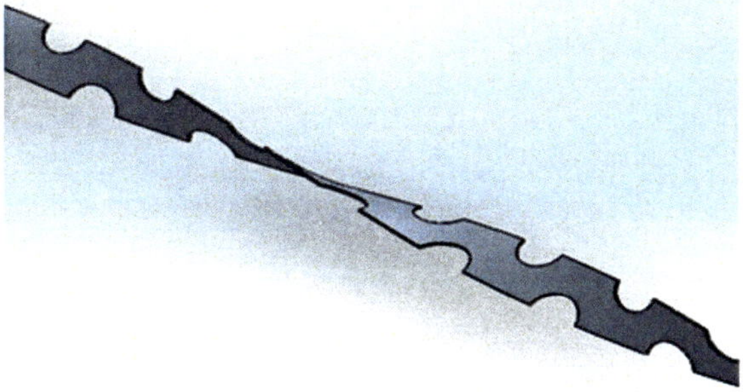

Fig. 2 Circular cutout

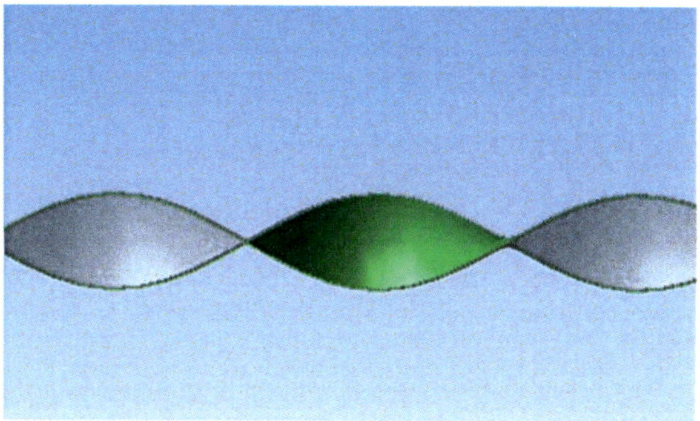

Fig. 3 Simple twited tape

4 Numerical modelling

The fluid flow is incompressible, three dimensional, turbulent and steady. The governing equations are as follows [28–31]:

(i) Continuity equation

$$\frac{\partial(\rho ui)}{\partial xi} = 0 \tag{1}$$

(ii) Momentum equation

$$\frac{\partial(uj\rho ui)}{\partial xi} = -\frac{\partial p}{\partial x} + \frac{\partial}{\partial xj}\left\{\mu\left[\left(\frac{\partial ui}{\partial xi}\right) - \mu uiuj\right]\right\} \tag{2}$$

(iii) Energy equation

$$\frac{\partial(\rho uiT)}{\partial xi} = \frac{\partial}{\partial xi}\left\{\left[\left(\frac{\mu}{Pr}\right) + \left(\frac{\mu t}{Prt}\right)\right]\left(\frac{\partial(\rho T)}{\partial xi}\right)\right\} \tag{3}$$

In the analysis, the turbulence model used was RNG k-ε model. There are two equations in this model for turbulence model (k) and the dissipation rate (ε) [32, 33].

$$\frac{\partial(\rho uik)}{\partial xi} = \frac{\partial}{\partial xj}\left\{\left[\left(\frac{\mu t}{\sigma k}\right) + \mu\right]\left(\frac{\partial k}{\partial xj}\right)\right\} + G_k - \rho\varepsilon \tag{4}$$

$$\frac{\partial(\rho ui\varepsilon)}{\partial xi} = \frac{\partial}{\partial xj}\left\{\left[\left[\left(\frac{\mu t}{\sigma k}\right) + \mu\right]\right]\right\}\left(\frac{\partial\varepsilon}{\partial xj}\right) + \left(\frac{\varepsilon}{k}\right)[C_{1\varepsilon}G_k - \rho C_{\varepsilon}\varepsilon] \tag{5}$$

$$\text{the generation of turbulent kinetic energy} = G_k = -\mu u_i u_j \frac{\partial(uj)}{\partial xi} \tag{6}$$

u = velocity of the fluid

μ = Dynamic viscosity of the fluid

μ_t = Turbulence viscosity = $C_\mu k^2/\varepsilon$

$$\text{The turbulence intensity I} = \frac{0.16}{\text{Re}1/8} \tag{7}$$

$$\text{Shear wall stress } \tau_w = C_f \rho u^2/2 \tag{8}$$

The mesh was generated, and the boundary conditions were applied. The flow at the inlet was assumed to be laminar with Re from 2300 to 40000 with corresponding inlet velocities 0.0940–0.8 m/s. With turbulence intensity 0.046 was considered. On the outer wall, a constant heat flux condition was imposed keeping in consideration that on a normal sunny day 1 kW/m^2 of energy is dissipated on any surface. No-slip condition was applied on tube wall and the twisted tape wall. In the case of solver setup, a SIMPLE algorithm was used along with second-order discretization for turbulent dissipation rates as well as turbulent kinetic energy. To study the dependence of the data acquired over the number of elements, grid dependence test was done with 46373, 92610, 142430 and 155404 elements for TCTT and 85899, 160824, 246706 and 283775 elements for CCT. The node independence test was carried on the CCTT and TCTT at Reynolds number 40000 (Table 1).

Table 1 Grid independence test for CCTT

No. of elements	Nu	%Change
85899	3.34476	–
160824	3.43476	2.69%
246706	3.42668	0.23%
283775	3.41959	0.20%

5 Result and Discussions

(a) **Data reduction and grid independence test.**

$$\text{Heat transfer (Q)} = \text{mass flow rate (m)} \times \text{Cp} \times (\text{To}-T_m) \quad (9)$$

where T_o is Initial temperature of water and T_m is mean bulk temperature

$$\text{Convective heat coefficient transfer, } h = Q_o/(T_w-T_m) \quad (10)$$

The data validation was done for all three cases with the help of grid independency test. The relation between the number of elements and % change in the Nu was considered for the grid independence test, and the relation can also be seen from Fig. 4. For carrying out the node independence test, the number of elements for the meshing was increased with the help of increasing relevance in the order of 40, 80 and 100. In the case of CCTT, the lowest % change in Nu was 0.009 for 142430 elements. Percentage change in Nu was least, i.e. 0.2% in the case of 283775 elements. The cases with the minimum % change in NU were considered.

(b) **Outlet temperature versus Reynolds number**

The outlet temperature versus Reynolds number relationship for STT, TCTT and CCTT is shown in Fig. 5. At Reynolds number 2300, the outlet temperature of the fluid in case of TCTT is 0.015% is greater than STT, whereas that of CCTT is 0.49% lower than that of STT. When the Reynolds number is further increased to 10000 the outlet temperature of water in TCTT and CCTT is 0.058% and 0.17% lower than STT. At Reynolds number 20000, the outlet temperature of water is 0.075% and 0.122% lower than STT in case of TCTT and CCTT, respectively. For further increase in Reynolds number to 30000 and 40000, the outlet temperature in case of TCTT is 0.077% and 0.075% lower than STT and 0.099% and in case of CCTT the difference is 0.099% and 0.083%, respectively.

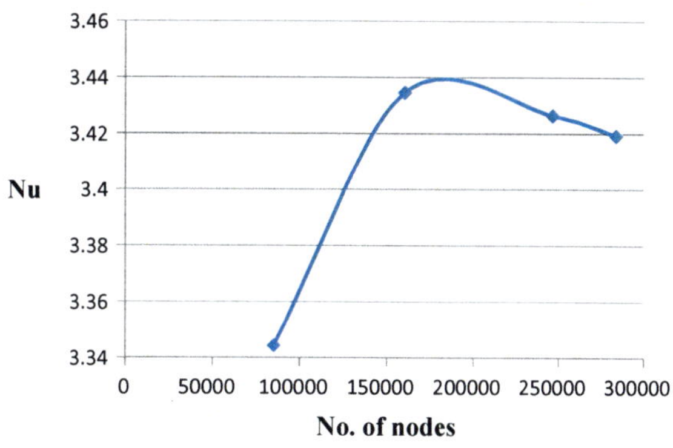

Fig. 4 Grid independence test for CCTT

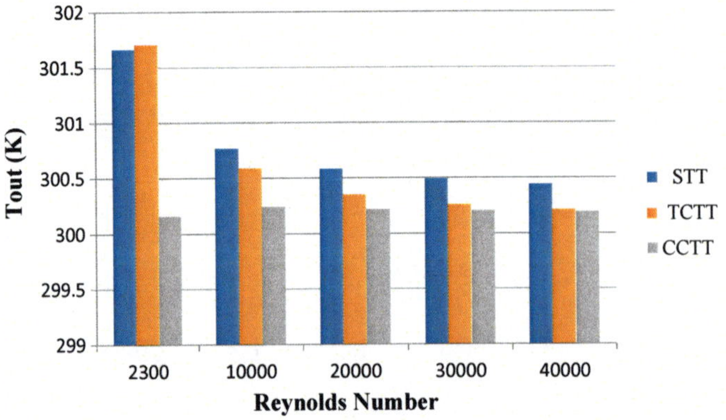

Fig. 5 Outlet Temperature versus Reynolds number

(c) Pressure drop versus Reynolds number

The pressure drop versus Reynolds number relationship for STT, TCTT and CCTT is shown in Fig. 6. At Reynolds number 2300 that is in laminar flow, the pressure drop in TCTT is 59% lower and in case of CCTT it is 35% greater than that in STT. When the Reynolds number is further increased to 10000 the pressure drop in STT is 60% and 45% greater than TCTT and CCTT, respectively. With further increase in Reynolds number to 20000, the pressure drop in case of STT is 62% and 68% greater than TCTT and CCTT, respectively. At Reynolds number 30000, the pressure drop in case of TCTT is 69% and CCTT is 77% less than STT. When Reynolds number is increased to 40000 the pressure drop in TCTT is 70% and in CCTT is 81% lower than STT.

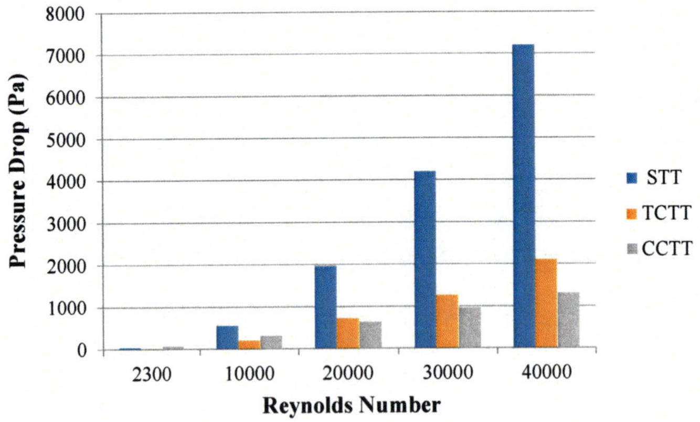

Fig. 6 Pressure drop versus Reynolds number

6 Conclusion

In this present study, the thermal performance of simple twisted tape, triangular cutout twisted tape and circular cutout twisted tape is numerically studied. The mains observations are

i The twisted tape is easy to manufacture and cheap accessory used in tubular heat exchangers and it helps in major improvement of heat transfer by increasing turbulence.
ii Inserting twisted tape inside a tube increases the turbulence intensity and also induces a swirl flow that enables heat transfer.
iii The outlet temperature observed was highest when the Re was in the laminar range. And with an increase in Re the outlet temperature decreases.
iv The pressure drop in simple twisted tape is maximum with respect to triangular cutout twisted tape and circular cutout twisted tape in the turbulent flow range.

References

1. K. Aroonrat, H.S. Ahn, D.-W. Jerng, L.G. Asirvatham, A.S. Dalklç, O. Mahian, S. Wongwises, Experimental study on evaporative heat transfer and pressure drop of r-134a in a horizontal dimpled tube. Int. J. Heat Mass Transf. **144**, (2019)
2. H. Allahyar, F. Hormozi, B. ZareNezhad, Experimental investigation on the thermal performance of a coiled heat exchanger using a new hybrid nanofluid. Exp. Thermal Fluid Sci. **76**, 324–329 (2016)
3. H.C. Anamika, Thakur, G. Nandan, A review on solar air heater performance using different artificial roughened rib, in *International Conference Emerging Development In Engineering and Technology (ICEDET-2016), Shivalik College of Engineering, Dehradun, India*, pp. 347–349, 2016
4. T. Agrawal, R. Ajitkumar, R. Prakash, G. Nandan, Sodium silicide as a hydrogen source for portable energy devices: a review. Mater. Today: Proc. **5**(2), 3563–3570 (2018)
5. M. Bahiraei, A. Godini, A. Shahsavar, Thermal and hydraulic characteristics of a minichannel heat exchanger operated with a non-newtonian hybrid nanofluid. J. Taiwan Inst. Chem. Eng. **84**, 149–161 (2018)
6. M. Bahiraei, N. Mazaheri, S.M. Hassanzamani, Efficacy of a new grapheneplatinum nanofluid in tubes fitted with single and twin twisted tapes regarding counter and co-swirling flows for efficient use of energy. Int. J. Mech. Sci. **150**, 290–303 (2019)
7. M.H. Bahmani, G. Sheikhzadeh, M. Zarringhalam, O.A. Akbari, A.A. Alrashed, G.A.S. Shabani, M. Goodarzi, Investigation of turbulent heat transfer and nanofluid flow in a double pipe heat exchanger. Adv. Powder Technol. **29**, 273–282 (2018)
8. S.W. Chang, K.-C. Yu, Heat transfer enhancement by spirally coiled spring inserts with and without segmental solid cords. Exp. Thermal Fluid Sci. **97**, 119–132 (2018)
9. E. Dabiri, F. Bahrami, S. Mohammadzadeh, Experimental investigation on turbulent convection heat transfer of SiC/w and MgO/w nanofluids in a circular tube under constant heat flux boundary condition. J. Therm. Anal. Calorim. **131**, 2243–2259 (2017)
10. A. Hasanpour, M. Farhadi, K. Sedighi, Experimental heat transfer and pressure drop study on typical, perforated, v-cut and u-cut twisted tapes in a helically corrugated heat exchanger. Int. Commun. Heat Mass Transfer **71**, 126–136 (2016)

11. F. Russo, N.T. Basse, Scaling of turbulence intensity for low-speed flow in smooth pipes. Flow Meas. Instrum. **52**, 101–114 (2016)
12. A. Hasanpour, M. Farhadi, K. Sedighi, A review study on twisted tape inserts on turbulent flow heat exchangers: The overall enhancement ratio criteria. Int. Commun. Heat Mass Transfer **55**, 53–62 (2014)
13. M.N. Labib, M.J. Nine, H. Afrianto, H. Chung, H. Jeong, Numerical investigation on effect of base fluids and hybrid nanofluid in forced convective heat transfer. Int. J. Thermal Sci. **71**, 163–171 (2013)
14. H. Bas, V. Ozceyhan, Heat transfer enhancement in a tube with twisted tape inserts placed separately from the tube wall. Exp. Thermal Fluid Sci. **41**, 51–58 (2012)
15. P. Promvonge, S. Eiamsa-ard, Heat transfer behaviors in a tube with combined conical-ring and twisted-tape insert. Int. Commun. Heat Mass Transfer **34**, 849–859 (2007)
16. A.M. Abed, H.S. Majdi, Z. Hussein, D. Fadhil, A. Abdulkadhim, Numerical analysis of flow and heat transfer enhancement in a horizontal pipe with p-TT and v-cut twisted tape. Case Studies Thermal Eng. **12**, 749–758 (2018)
17. B. Salam, S. Biswas, S. Saha, M.M.K. Bhuiya, Heat transfer enhancement in a tube using rectangular-cut twisted tape insert. Procedia Eng. **14**, 96–103 (2013)
18. C. Thianponga, P. Eiamsa-ard, P. Promvongea, S. Eiamsa-ard, Effect of perforated twisted-tapes with parallel wings on heat transfer enhancement in a heat exchanger tube. Energy Procedia, 1111–1120 (2012)
19. M. Farnam, M. Khoshvaght-Aliabadi, M. Asadollahzadeh, Heat transfer intensification of agitated u-tube heat exchanger using twisted-tube and twisted-tape as passive techniques. Chem. Eng. Proces. Process Intensification **133**, 137–147 (2018)
20. S. Eiamsa-ard, P. Promvonge, Performance assessment in a heat exchanger tube with alternate clockwise and counter-clockwise twisted-tape inserts. Int. J. Heat Mass Transf. **53**, 1364–1372 (2010)
21. B. Kumar, M. Kumar, A.K. Patil, S. Jain, Effect of v cut in perforated twisted tape insert on heat transfer and fluid flow behavior of tube flow: an experimental study. Exp. Heat Transfer. **32**, 524–544 (2018)
22. S. Shabanian, M. Rahimi, M. Shahhosseini, A. Alsairafi, CFD and experimental studies on heat transfer enhancement in an air cooler equipped with different tube inserts. Int. Commun. Heat Mass Transfer **38**, 383–390 (2011)
23. S. Eiamsa-ard, P. Seemawute, K. Wongcharee, Influences of peripherally-cut twisted tape insert on heat transfer and thermal performance characteristics in laminar and turbulent tube flows. Exp. Thermal Fluid Sci. **34**, 711–719 (2010)
24. S. Zhang, L. Lu, C. Dong, S.H. Cha, Performance evaluation of a double-pipe heat exchanger fitted with self-rotating twisted tapes. Appl. Therm. Eng. **158**, (2019)
25. N. Mashoofi, S. Pourahmad, S. Pesteei, Study the effect of axially perforated twisted tapes on the thermal performance enhancement factor of a double tube heat exchanger. Case Studies Thermal Eng. **10**, 161–168 (2017)
26. S. Pourahmad, S. Pesteei, Effectiveness-NTU analyses in a double tube heat exchanger equipped with wavy strip considering various angles. Energy Conver. Manag. **123**, 462–469 (2016)
27. P. Naphon, Heat transfer and pressure drop in the horizontal double pipes with and without twisted tape insert. Int. Commun. Heat Mass Transfer **33**, 166–175 (2006)
28. P. Li, Z. Liu, W. Liu, G. Chen, Numerical study on heat transfer enhancement characteristics of tube inserted with centrally hollow narrow twisted tapes. Int. J. Heat Mass Transf. **88**, 481–491 (2015)
29. A. Fan, J. Deng, J. Guo, W. Liu, A numerical study on thermo-hydraulic characteristics of turbulent flow in a circular tube fitted with conical strip inserts. Appl. Therm. Eng. **31**, 2819–2828 (2011)
30. R. Kumar, G. Nandan, G. Dwivedi, A.K. Shukla, R. Shrivastava, Modeling of triangular perforated twisted tape with V-Cuts in double pipe heat exchanger, in *Materials Today: Proceedings* (2020)

31. G. Nandan, N.A. Wani, Numerical study of copper oxide/therminol vp-1 nanofluid in solar parabolic trough collector, in *Advances in Engineering Design*, vol 1 (2021)
32. M.Z. Hayat, P. Tanwar, H. Chauhan, G. Nandan, R. Shrivastava, Numerical analysis of double pipe heat exchanger using cross twisted tape with v-cuts, in *Lecture Notes in Mechanical Engineering* (Springer, Singapore, 2020)
33. M.Z. Hayat, G. Nandan, A.K. Tiwari, S.K. Sharma, R. Shrivastava, A.K. Singh, Numerical study on heat transfer enhancement using twisted tape with trapezoidal ribs in an internal flow, in *Materials Today: Proceedings* (2020, Oct)
34. G. Nandan, N.A. Wani, S. Kumar, Augmentation of thermal performances in a tubular heat exchanger using internal flow baffler: recent study, in *Advances in Enginering Design*, vol 1 (2021)